Consulting by Legends Library

UM logo is a registered trademark in USA and is pending.

ISBN: 978-1-944200-93-0
1. Science 2. Earth Science 3. Life Science 4. Space Science 5. Natural Science

For permissions email:
support@universalmodel.com

For information or to order more copies please go to:

UniversalModel.com

SUMMARY

AN INTRODUCTION TO THE
UNIVERSAL MODEL
A NEW MILLENNIAL SCIENCE

Brooke E. McKay

Foreword by Dean W. Sessions

Edited by Russell H. Barlow

Foreword

The UM Summary, written by my daughter, Brooke E. McKay, is an Introduction to the *Universal Model, a New Millennial Science*. I enjoyed watching each of our five children grow up learning about the UM while experiencing our many family adventures across the United States as we studied various aspects of Nature. Now, as parents themselves, our children turn to the UM to teach their own children about the wonders of our Universe. Indeed, growing up with the UM has given Brooke a unique perspective with which to write this summary.

Drawing much of her text directly from the UM, Brooke introduces each topic in a simple and understandable format. As one of our premier teachers and instructors, she possesses a unique understanding of the UM and how it can affect our lives. Brooke's motivation to write this book has come from her desire to share and teach the UM to others. Homeschooling her four children has presented her with opportunities to teach by example as she demonstrates simply the principles and laws found in the Universal Model.

It is hard to describe my joy as I observe our children and grandchildren learning how Nature works. Thank you, Brooke, for helping us all to more easily understand and comprehend the UM's new discoveries.

Dean W. Sessions
Author of the Universal Model

April 2017

Preface

The Universal Model has always been a family journey. Asking fundamental questions, making incredible discoveries and sharing the truth of these discoveries with others were typical topics of conversation around our family dinner table. It started almost three decades ago when my father first caught the science bug. From then on we spent most weekends and vacations knee deep in "research and development," or as my siblings and I liked to called it, "family vacations to every national park and rock store in the country."

We saw it all as we hunted for arrowheads, rock-hounded for crystals, combed the desert floor for surface chalcedony, or broke open shale slices searching for fish fossils. We visited biospheres and museums, collected sand samples from every beach and dune, and stopped to take photos of literally thousands of rocks exposed in highway road cuts. Each place we stopped was "The coolest place ever." And each memory captured in a photo for the family scrapbook, also became a new discovery for the Universal Model.

I was taught the new discoveries found in the UM pretty much my entire life. It was a struggle to sit through my public school science classes, knowing the information being taught contradicted directly the truths that my Father had shown me. I grew up seeing first hand that fossils did not take millions of years to form because I watched my dad make one in our garage in just days. I understood that the huge Arizona Meteor Crater we visited could not have been made by a meteorite. I knew there was no way the interior of the Earth was full of magma. I also knew that the entire Earth was covered by a global flood around 4,360 years ago, and that there is actual, observable scientific evidence proving it all.

While we each have a unique story to tell, filled with our own personal experiences, my story involves helping the world rediscover scientific truth; truth either swept under the proverbial modern science rug or perhaps just waiting for its discovery.

There are many discoveries in the Universal Model; true scientific discoveries that offer hope to some that there are actual answers to their questions about the way Nature works, not just unfounded theories that are now being taught as fact. The UM is huge, not only in its impact to the world of modern science, but also in its physical form. There are over two thousand pages in 30 plus chapters with hundreds of diagrams and photos… it is a lot to take in and the reason for this summary, or introduction book.

Inside this book you will find many of the fundamental questions that started it all. Some of these questions my father had asked many scientists and professors around the world; questions that no one seemed to have answers for. The Universal Model gives these answers, and many of them appear in this introductory book. Extraordinary claims accompany most of the answers in the UM, and they often run contrary to what modern science currently teaches as fact in the world's classrooms. These extraordinary claims demand extraordinary evidence, which the reader will find in the UM, but which I omit here in order to keep this book brief.

Be prepared to fall in love with science all over again, just like you did the first time you added baking soda and vinegar together and made an exploding volcano, or grew rocks out of molten magma… wait a second… has anyone actually done that or seen it happening in Nature? Nope! Because magma isn't part of the rock-making process. Science is amazing and you don't have to be a kid to find the joy in learning true science!

It is my intention that by seeing the UM's big picture in a condensed form that you will more easily understand how encompassing and truly significant the new discoveries found in the UM are, and how they will affect each field in modern science. I hope you enjoy reading the UM Summary as much as I enjoyed writing it, and I hope that it leads you to search for the scientific truth out there awaiting to be found. Happy discovering!

Brooke E. McKay

April 2017

Table of Contents

Chapter 1
The Universal Model
A New Millennial Science

*"This magnitude of new scientific discovery
has **never** been presented to the public before."*

Universal Model, Subchapter 28.1

Have you ever gazed at the nighttime sky and marveled at the Universe, pondering how big it really is or how it all came to be? Have you ever been relaxing on a warm beach with your toes in the sand and wondered where it all came from and why each little grain is similar in size? In moments like these have you ever wished that time would slow down or speed up or perhaps wondered how time even exists?

These are questions brought on by the sometimes puzzling reality of Nature and its workings. The purpose of science is to describe and explain Nature so that you can more easily understand and comprehend it. However, the current perception of science is that it is anything but easily understood.

The questions asked above, believe it or not, have very simple answers. These answers and hundreds more are found in a new book of scientific discovery called the Universal Model – A New Millennial Science. Science today, identified throughout the Universal Model as **modern science**, does not have the complete knowledge needed to answer these questions, nor the wisdom to explain or comprehend them. This realization is why the Universal Model came to be. The purpose of the Universal Model is to restore truth and order in science by identifying new natural laws. Most natural laws are simple to understand, easily tested, and repeated by the public, not just by educated, professional scientists. This mindset allows everyone the opportunity to discover the answers to how Nature works.

What is the UM?

Short for *Universal Model – A New Millennial Science*, the UM consists of a three volume book containing thousands of images and science journal references. The Universal Model is a living document, full of new scientific truths that have been discovered over the last few decades and it is the product of ongoing, continuous research and discovery. Written for the typical individual to enable a better understanding of the world around us, the UM comprises the largest single body of *new scientific discovery* ever written. It contains hundreds of new discoveries gathered from across all the natural sciences.

Who developed the UM?

The Universal Model is a collaborative research project developed by private individuals and written by the author, Dean W. Sessions. The UM exists independent of any educational, government, or modern science institution. These organizations tend to favor the modern science 'establishment' and by not attaching this new scientific work to any group, the UM was able to maintain objectivity and independence in its discovery of new scientific truth.

The new discoveries published in the UM were accomplished independent of any modern scientists' direct involvement. Credentialed scientists tend to specialize in only one field of study, but the UM discoveries came about in part because the author, a generalist, was able to study and simultaneously apply research from multiple fields of science. This proved important because the answers to the biggest questions often came from completely different areas of study. For example, in the UM you will learn how questions posed in the field of biology find answers from geology. The UM demonstrates the interconnectivity among all of the sciences, and how each fits like a piece of a puzzle.

For whom was the UM written?

The Universal Model is for *everyone*. Although it is written on a college-entry level, the concepts are simple enough to teach grade-school students. Contrary to general thoughts regarding modern science, the UM science is neither complex nor hard to understand. People of all ages quickly and easily grasp its principles and concepts because the they are based entirely upon testable and repeatable new natural laws, and Nature in and of itself is simple. You can verify many of the natural laws by personally observing Nature or by experimentation in your own home.

Where was the UM written?

Many of the discoveries and most of the writing of the UM took place in the state of Arizona. It is unique in the sense that it is home to some of the largest and most incredible geological sites in the world. These sites proved crucial in the development of the UM and its discoveries. Arizona has more than 300 clear, sunny days a year enabling the capture of hundreds of astronomical photos and other images included in the book, some of which provide support for the extraordinary claims throughout the UM.

Every year, Tucson, Arizona hosts the world's largest rock, mineral, and fossil show, drawing tens-of-thousands of spectators, enthusiasts, and vendors from around the world. Specimens obtained or discussed at the rock show helped shape the UM's new geological view. Arizona is also home to the world's largest petrified forest, the largest most studied 'confirmed meteor' crater in the world, and one of the world's largest canyons; each a profoundly influential geological site in the development of the UM. Arizona is truly the heartland of the UM.

When was the UM written?

The author of the UM began official research for the Universal Model in 1990 after asking fundamental questions related to the fields of archeology and geology for which he was not able to find the answers in either current science books, or from educated scientists and professors. The writing of the UM began in earnest around 2000, but research, words, and chapters continue to accumulate today. The website, **UniversalModel.com** was launched publically on October 15th 2016. There, you can gain access to the UM's chapters, view UM videos, blogs, reviews, and begin your journey in discovering a New Millennial Science.

How can we trust the credibility of a new science book that is not written by credentialed scientists?

A scientific degree is not necessary to learn or discover scientific truths. The new natural laws and many discoveries in the UM prove this fact. If the new discoveries identified in the UM demonstrate repeatability through experimentation and stand the test of time, it does not matter who discovered them.

The UM encourages everyone to read and discover truth for themselves. We believe true science is simple and enjoyable to learn. Simple enough to test the discoveries yourself, or to see the evidence with your own eyes and enjoyable once you realize the laws and discoveries in the UM actually make sense. This, along with the fact that the UM provides answers to so many of

Nature's greatest questions, backed by empirical evidence, stands in contrast to some groups who offer theories with no real scientific evidence to support them.

Even with direct physical evidence, the UM will not likely change the mind of *every* modern scientist, especially those who believe so strongly in their theories. Still, if the public will begin to question scientists objectively, requiring the physical evidence for the theories that they insist are scientific fact, then the truth will soon come forward. With the public educated and armed with observable, testable evidence, we no longer need to accept the "trust me" mantra of the world's renowned scientists.

Is the UM a Creationist publication?

For many years, creationists and evolutionists have engaged in an ongoing battle, both presenting insufficient, untestable scientific evidence to justify their beliefs. While no creationist group funded the development of the Universal Model, many of the new truths outlined in the UM constitute physical scientific evidence in support of beliefs expressed by creationist groups and other religious organizations. While the Universal Model includes physical scientific evidence for the Earth's formation, or creation, it presents a completely new meaning to the word 'creation' itself. The UM is revolutionary in its scope due to the large amount of evidence presented, including elements of 'the creation story' told here for the very first time.

Why is the UM book so large and written like a textbook?

Few people read large textbooks just for fun, but the UM is a very unique type of textbook. Its writing and exposition made possible the birth of Millennial Science, and it required a large number of pages to express the myriad of extraordinary claims, discoveries and evidences required to replace modern science with Millennial Science.

What will we find on the UM's pages?

Each chapter includes detail sufficient to stand on its own; however, by reading the chapters together, they build on previous chapters, unfolding beautifully connected pieces of Nature's puzzle; one that is easy to grasp and exciting to comprehend. One of the ways in which the UM demonstrates its new science principles is by including hundreds of photos and diagrams providing visual aids to supplement the learning process.

To establish modern science's beliefs and ideas, or to present their discoveries and direct observations, UM researchers poured through thousands of articles from scholars and professionals in all natural science fields. Many of their quotes made their way into the UM and are highlighted in **blue** for easy recognition. These quotes and citations come from well-known and prestigious scientific journals, magazines, and from hundreds of books and other online articles.

The first four chapters in the UM include the Introduction chapters, extremely important prerequisites to the physical science chapters that come after them. They contain principles that set the foundation for comprehending the new discoveries that follow.

The full scope of the project includes three primary systems, each uniquely connected to the others. These systems include the Earth System, the Living System, and the Universe System. Each one encompasses different aspects of Nature, covering many thought provoking questions along with powerful answers to those questions, many of which have never before been known. The new discoveries in these systems dramatically change how we view science, Nature, and what we think we know about the world and Universe around us. The systems each comprise one volume, which is divided into numbered chapters and subchapters. Throughout the book, references are given to help the reader confirm the information for themselves.

The final UM chapters take a look at the future of science, with respect to how new UM discoveries might affect the world. These Futurity chapters also discuss the impact the Universal Model will have on Millennial Science and Millennial Education in the years to follow.

Why do we need a New Millennial Science?

The Universal Model recognizes that many aspects of science and discovery that have occurred over the past several centuries are true. The temperature and pressure at which water freezes, the existence of cells, the discovery of DNA in all living organisms and many other accurate facts of science provide us solid groundwork from which to approach Nature. Because these are *not* in question, they are *not* a part of the science addressed in the UM.

What the UM does address and what is of primary concern are the scientific *theories* that modern science increasingly teaches as fact, even though they remain unproven. These unproven theories form the foundation of much of modern science today. New discoveries in the UM shine a bright light on these false theories, causing them to give way to new natural laws that form the foundation for a New Millennial Science. Built on the Universal Model, Millennial Science will continue the discovery of other natural laws while it continues to develop a map of nature that correctly describes the Universe in which we live. More details concerning the need for a New Millennial Science are covered in chapters 2 and 7 in this book.

Why are natural laws so important?

As previously stated, the purpose of the UM is to restore truth and order in science by identifying new natural laws. A natural law is simply a statement of truth. Natural laws are generally simple and often testable by the non-professional scientist. Natural laws exhibit a cause and effect relationship; they have a predictable consequence from a stated action.

Natural laws help describe and explain Nature and should form the foundation for all science. Because natural laws outlast theories, they are the pinnacle of scientific discovery.

Why are true definitions so important?

To communicate an idea effectively, it is important to establish common definitions for the words we use. The UM acknowledges the value in this and defines important words, less common scientific words, and words that have unclear or contradictory meanings. By using common definitions, everyone can better understand what the author intends to convey. Conversely, incorrect definitions can cause confusion and disorder. No agreement can be reached if definitions are not correct. As you read the Universal Model, you will see that many incorrectly defined science words, including the word science itself, have inhibited humanity's understanding of Nature's truths for many years.

Why is performing experiments so important?

Experiments and science go hand in hand. However, some modern scientists believe that one can be a scientist today but not perform any experiments. The UM will show you that this is a false ideology and that experimentation must be present for science to actually be true science. There is no better way to learn than from direct, hands-on experience. You are encouraged to perform as many of the experiments in the UM as you can. Most of them are simple enough to perform at home and will contribute to a deeper understanding and comprehension of the concepts presented.

After reading the UM, performing the experiments, and making your own observations, you will have a more accurate perspective with which to judge the work. You can frame your own questions as you make your own discoveries; the world is still full of things awaiting discovery! It is our hope that with the new principles and discoveries found in the UM, others will continue to ask the questions that will move the world forward towards even greater knowledge, wisdom, and discovery in Nature.

Why is asking questions so important?

Questions stimulate the human mind. They initiate inward reflection and outward expression. Most importantly, answers come by asking questions. Questions are the foundation of learning and there is much to be said of this principle. Without the principle of questioning everything with an open mind, the UM would not exist.

In the UM, a question about a basic concept in Nature is a **Fundamental Question** or **FQ**. An answer to a question that provides a fundamental understanding of Nature is a **Fundamental Answer** or **FA**. Highlighted green in the text throughout the Universal Model, the use of FQ's and FA's aid in the discovery of Nature's secrets. While there will be many questions that seem completely new; you may have already asked some of them yourself. These are questions that in many instances, modern science has not, or is currently unable to answer. Although questions come in many forms, they can be simplified into one of six different types of questions which are found in the new UM Learning Process.

Why do we need a new Learning Process?

The development of the Universal Model fostered the introduction of a new learning process, which included the recognition of a true definition of the words knowledge and wisdom and how they differ. Knowledge provides understanding, which comes from answers to the questions **what, where, when** and **who**. Wisdom provides comprehension and comes from answers to the questions **how** and **why**. An increase in our understanding and an expansion of our ability to comprehend allows us to describe and explain Nature more completely, which is exactly what science is all about!

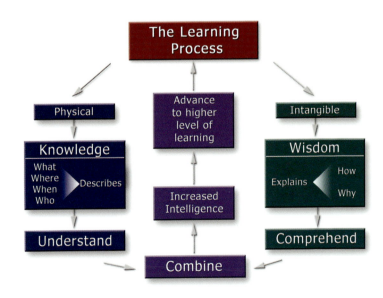

Chapter 2
The Big Picture
Why we need a New Millennial Science

"Without truth in science there is no natural law.
Without natural law, there is no science."

Universal Model, Subchapter 3.4

Scientists today have the honor, as well as the burden, of being known as the expert when it comes to explaining how Nature works. With this, comes the inherent obligation of truthful discovery and factual knowledge. In our search for understanding we tend to trust the words of the professionals in whatever field we are inquiring. Their opinions provide direction in our search for knowledge and we are told to trust that the information they teach is correct.

One of the greatest principles prescribed throughout the UM is, "Question everything with an open mind." While this might seem contrary in a world where you are always told to just trust the professional, it is how truth is found. By allowing ourselves to open up to the thought that we do not know everything there is to know, and by learning to ask questions to find out those things we may not know, we open ourselves up to the discovery of greater knowledge. This concept challenges us to answer this very important fundamental question:

What if what we believed to be true, simply is not?

If someone showed you a dime and asked its value, you would likely answer ten cents. However, if you were given additional information about that dime, such as its mint date of 1950, your answer may change dramatically. Dimes minted in 1950 were 90% silver, making its silver-content value considerably higher than its face value, making the true worth of that dime range from 1 to 90 dollars. The open minded consideration of new knowledge or information allows us to learn a true understanding of the dime's worth. Likewise, with the new information, observations, and discoveries, (or new knowledge) incorporated in the Universal Model, you will gain a more correct understanding of the Universe, the Earth, and even yourself.

It is a true fact that there are many things modern science does not know. This by itself would not create a problem if scientists actively used a Universal Scientific Method in their quest to discover and learn truths that others can easily test and reproduce.

Someone once said that the greatest discovery in science was the discovery of the scientific method of discovery. Say that five times fast. So, what is the *definition* of the greatest discovery in science? There are actually countless versions of the scientific method, which should raise a red flag to anyone involved in the scientific discovery process.

Is there a standard Scientific Method?

The scientific method discussed in textbooks and endless online sources appears in countless variations. Some go so far as to say that such a method doesn't even exist. It is clear to anyone who does a quick search that no standardized, single set of steps defines the modern 'Scientific Method'. Granted, there are common elements among the various methods, such as theory, hypothesis, or observation, however, there is little if any consensus, even within science fields, regarding how or when to apply certain elements.

How are educators and parents supposed to teach a standardized method of scientific discovery if science itself does not have one? Equally important, how do we hold scientists accountable for the accuracy and reproducibility of their experiments if there is no Universal Scientific Method (USM) by which to measure and judge their work?

Is uniformity in our approach to science THAT important? Are the steps we use or how we apply them critical? Can we change the order or delete certain steps of the method and still get the same result? You would scoff at a contractor who began building a home starting with the roof and ending with the foundation, or decided indoor plumbing just isn't important, and left it out. Yet, this is what we see in many versions of the scientific method; certain essential parts left out, such as natural law, while theory occupies the final step, the endpoint of scientific discovery.

To establish order in science we **must** establish a method to which everyone must adhere. Allowing every scientist to make up and employ their own version of the scientific method is just as chaotic as it would be to let accountants make up their own tax forms or lawyers to make up their own set of rules in a courtroom.

Standardization is necessary for order. We understand that no one wants someone or some method to stifle their creativity and imagination, however, true science is about discovering truth in Nature and for that to happen, the public must require that scientists adhere to a standardized universal scientific method to ensure accuracy and accountability.

The UM understands that scientific truth and reason *can* and will explain the Universe simply, and that there *must* be a method that all can employ. For the very first time, the Universal Model establishes a Universal Scientific Method which was the framework used during UM research projects and experiments and allowed hundreds of extraordinary discoveries and dozens of new natural laws throughout every major field of science to be found. Something that no other single work of science has ever done.

Are there more scientists today than there were a hundred years ago? Are we spending more research money today on science than we did a hundred years ago? Do we have more advanced technology to aid in that research than we did a hundred years ago? The answer to all three of these questions is a resounding, **YES**. This stimulated a mind blowing fundamental question for UM researchers:

Why has modern science discovered less natural law today than it did a century ago?

Even though there are many more scientists with billions of dollars spent annually on scientific research and tremendous advances in technology, there has been a noticeable decline in natural law during the past century. In fact, modern science has not posted *even one new significant natural law* over the past 100-year period. The lack of standardization in how we make and record scientific discoveries has certainly contributed to the decline in natural law, however UM researchers found another compelling catalyst leading to the last centuries' lack of laws.

From around 500 BC to the mid-1800s, scientists understood that if they proposed a theory that their physical and direct observations and evidences were not able to prove, they discarded or revised their theory. Beginning around the 1900s, a new trend emerged as scientists stopped practicing this principle, eventually deciding that *their theories were the endpoints of science,* no matter what their observations or evidence showed. This is why we need a New Millennial Science.

In the UM, when the truth of a theory cannot be demonstrated, and when that theory is taught as fact, we call it a **Pseudotheory**. Pseudo, a Greek prefix attached to words to indicate false or unreal, describes a number of modern science theories explained throughout the UM. They are taught as fact worldwide, sometimes even introduced to preschool aged children. The UM gives a brief introduction on how each pseudotheory is taught and then, the pseudotheory is examined by asking important fundamental questions. If the theory does not stand up to the scrutiny of the fundamental questions, it is replaced with true models and, in some cases, new natural law.

What is the difference between theory and natural law?

In modern science, the words theory and law are often used interchangeably, giving the impression that they have the same meaning. When we demote natural law to theory or when we place theory on equal status with natural law, we violate the maxims of truth.

One real problem that occurs when theories are taught as law is that people come to believe that the theories are 'empirically' true, meaning verifiable by experience or experiment. For something to be empirically true, it must be observable, demonstrable, or repeatable, and many modern science theories are not.

What is the result of a century devoid of new significant natural laws?

The UM provides answers as to why modern science is no longer discovering any new significant natural laws. The primary reason is that they are not even looking for natural laws as they claim their theories are the endpoints of science. Because of this, the Universal Model dubs the last hundred years or so of scientific inquiry as the Scientific Dark Age. Few scientists are aware of the Scientific Dark Age, and even fewer know why it exists. Of the hundreds of extraordinary claims made in the UM, there is probably none so bold as to claim that modern science exists in a dark age today and has for more than a century. There is a crisis in the scientific community, even though most researchers have entered their fields thinking all is well. But when science goes for decades without the discovery of any significant new natural laws, there is an obvious problem.

What is the Big Picture of Modern Science?

Even after viewing each individual piece of the modern science belief system, it can be difficult to understand their big picture. The answer to this question is better understood by viewing Nature as a whole, which will clarify the controversial nature and illogical foundation of the modern science story. In science today, as pieces of the big picture are researched, taught and reported on in science magazines, journals and the media, you rarely hear of the 'Big Picture' because it just sounds too ridiculous, too unachievable. Despite how ridiculous it may sound, scientists try continuously to describe and defend the modern science big picture.

The Big Picture of Modern Science says essentially that humans evolved from an ape-like common ancestor and apes evolved from bacteria, bacteria appeared out of chemicals, chemicals came from the Big Bang, and... the Big

THE BIG PICTURE OF MODERN SCIENCE

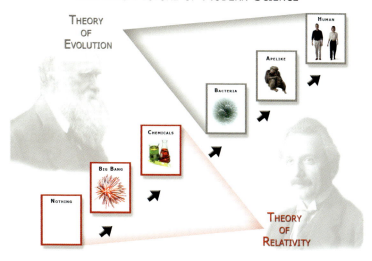

THEORY
OF
EVOLUTION

HUMAN

APELIKE

BACTERIA

CHEMICALS

BIG BANG

NOTHING

THEORY
OF
RELATIVITY

Bang came from nothing. *Yes, everything, including yourself, came from **nothing**.*

Throughout the UM you will read quotes from scientists who agree with this concept. However, after seeing this big picture, it should be no mystery as to why modern science remains unpopular with the masses. Not only is it complicated, it seems silly to imagine such a chain of events actually occurring, and it does not represent the consensus of the majority of the people of the world.

How will the UM end the Scientific Dark Age crisis?

There may never have been a ***bigger crisis*** in science than right now. Billions of taxpayer's dollars have and are being spent in an attempt to discover what happened at the very first moment, when everything, including you, appeared from nothing. The Universal Model is the answer to this crisis with its crucial experiments and observations available to the public, as well as the scientist. Each person can judge the new scientific evidence contained in the UM for themselves by evaluating the new data, observing the new natural laws and comparing the old ideas to new principles.

Importantly, while the general intent of the UM deals with correcting errors in the foundations of modern science, *not all science is based on false theory*. There is much in science today that comes from established correct principles and natural law. That which is true will continue to stand the test of time and will remain a viable part of scientific knowledge.

True science describes and explains Nature using natural law. In contrast, modern science has fallen into a dark age during the last century by relying on scientific theories unable to correctly describe and explain Nature. This led to the mingling of technological success with scientific theory, lending an air of credibility as compensation for failures in science.

Without doubt, the UM's primary, extraordinary declaration consists of both a claim and a discovery. All the UM evidences lead to the same conclusion: *The scientific establishment has languished in a Scientific Dark Age over the past century and the UM is destined to end it, ushering in a new Millennial Science.*

Chapter 3
Restoring Truth & Order
Identifying New Natural Laws

"No one can change the way Nature works, we can only observe what is, was and will be - which is truth."

Universal Model, Subchapter 27.4

There is order in Nature, which at first may not be apparent. The stars and galaxies appear as though they were carelessly strewn through the heavens and rivers seem to meander aimlessly across the countryside. Sand and dirt appear to lay randomly sprinkled around and plants and animals seem to be scattered throughout the Earth with no real purpose. However, as you shall soon read in the UM, there is order, direction and purpose throughout our Universe. And until now, this state of order has remained largely unrecognized in modern science.

It is a goal of the Universal Model to discover many of Nature's most basic, simple truths hidden from our view, and then to assemble these pieces of truth in their proper order.

Why does modern science's view differ so greatly from Nature's reality?

In the following comparison, two puzzles assembled using the same pieces depict two very different points of view. The pieces of the puzzle represent the processes, laws, and observations of Nature, which includes the Universe and its workings.

Modern Science Puzzle   **Nature's Puzzle**

The puzzle on the left describes how modern science views Nature; disconnected, confused, and chaotic with pieces forced where they do not belong. Modern science maintains theories that are simply not true resulting in this chaotic view. The puzzle on the right illustrates Nature's true beauty. This arrangement renders an understandable and cohesive image because answers are not forced to fit false theories. Truth is simple, and when truth is found in Nature it makes understanding Nature simple, unfolding to our eyes the intentional order of our Universe and the world around us.

Who can assemble the pieces of Nature's puzzle?

Some think science is only for the professionally educated. Some scientists have taken it upon themselves to decide that they alone can understand and comprehend Nature. Others believe that there are few who can actually learn the mysteries of the Universe because it is just too complicated. In reality, true science is simple enough for *everyone*. More than ever before, science is for the common person and the whole of society. In fact, many historically significant discoveries came not from scientists as you might think of them today, but rather by individuals who simply had a great desire to learn and discover truth.

Why is truth so important?

Truth, like knowledge and wisdom, must be clearly defined for real progress to occur. Furthermore, the definitions of all words that are based on truth demand that the definition of truth, also be true. In essence, truth is the foundation upon which all else should be built and as such it deserves our attention in defining it. The UM details the essential role truth plays in our daily lives. You will read that defining truth involves more than just a general description of the word. In science, when a statement accurately predicts an effect over a continuous period of time, we consider that statement true. Thus, with the fundamental test of time, we can define truth.

There are truths in science that are self-evident. For example: a living human body contains water, a solid cube has six sides, and the order of color in a natural rainbow remains the same. Each of these statements is true because they state what is right now, what was, and what will be as far as we know.

The pseudotheory chapters in the UM demonstrate where truth has been mingled with error. We establish truth through observation and verification. The UM acknowledges that many scientists and researchers have made significant observations and identified many truths in the last century, but when those truths do not fit with current popular theory, they are molded and changed, with parts of the truth ignored so that it can better match or support current theory. Thus, favored pseudotheories continue to be taught to students year after year.

In fact, the words truth and law have actually been eliminated from the definition of science in some recent, post 1990's encyclopedias. Many find it disheartening that truth must be defended in science today. While not all scientists believe this way, many across the world teach and believe that truth is unknowable. Examples of how science denies truth are evident throughout the Universal Model.

What has happened to the idea of discovering truth?

Those who are truthseekers often end up being the men and women who make some of the most important scientific discoveries, with or without college degrees. The same holds true for those involved in the Universal Model.

My father often relates a story from a time when he first began his UM journey. A family friend who is a physics professor taught him something of great importance when he asked this friend what someone should do if they felt that they had discovered something of importance in a particular field of science, but did not have the requisite credentials in that field. He thought his friend would advise him to pursue a degree before publishing, but was surprised by his answer: *"If what you have found is important and true, it doesn't matter where it comes from, publish it."* My father learned a valuable lesson; **truth is independent of its origin**.

When truth is restored to many aspects of science in our day, it will come with the discovery of new scientific truths. The Universal Model will demonstrate the existence of truth, by the discovery of truth.

Chapter 4
The Earth System
A True Understanding of Our Earth

"We must leave the realm of scientific uncertainty and move on
to discover nature's *geo-logical* laws and scientific *certainties*...
Nature is too beautiful to not understand."

The Universal Model, Subchapter 6.13

A desire to be understood runs deeply through us all. The Earth itself also begs to be understood, leaving thousands of clues so that we can discover its deepest secrets, most of them right in front of us. Pure, white sand beaches and majestic grand canyons, brightly colored gems and wispy green auroras, acres of rainbow colored petrified wood and mile-wide craters. The Earth tells its own story through these physical evidences.

We find a majority of these evidences on the Earth's surface and modern science, to this day, does not have a true understanding of how they got there or how they formed. By using the Universal Scientific Method and with new information, observations, discoveries, including new natural laws, the Earth System will offer you a true understanding of the origin of our Earth and the geological beauty that surrounds us.

Asking fundamental questions and gaining a true understanding of the Earth's geology will help you answer the questions of what, where, and when, such as: What is quartz? Where does the Earth's heat originate? When do volcanoes erupt? The answers to questions like these help you understand the Earth's processes and provides you the opportunity to explain how and why they happen. The Universal Model considers current scientific data and introduces new scientific evidence, creating a new geological understanding which allows the recognition of the current theories' errors while laying the foundation for the Earth's true geology.

Modern geology relies on theories built upon theories, making the field of geology 'geo-theoretical' instead of geo-logical. Without observation, a theory is merely philosophy, not science. The Earth System covers many of these theories in depth, asking fundamental questions that modern geology has left unanswered, or has never even bothered to ask.

When a pseudotheory—a false theory being taught as fact—has been taught for over 100 years, it becomes ingrained in our collective thoughts and it becomes difficult to see outside the self-created box. Such is the case with modern geology. Some of the most fundamental questions that the UM asks remain unanswered by current theory, greatly diminishing our ability to understand our Earth.

The Magma Pseudotheory - UM Chapter 5

One of these pseudotheories, taught even earlier than elementary school, is the Magma Pseudotheory, described and debunked with physical and observable evidence in chapter 5 of the Universal Model.

You are, or were, taught in school that the Earth is a sphere because it was once a molten ball of liquid rock, and that the interior of the Earth is still molten magma. However, geologists have long known that there are major problems with this theory, yet they rarely talk about it. The UM contains many quotes from scientists themselves admitting to their confusion. The questions of where magma originates and how it is generated remains speculative, yet the theory continues to be taught as fact.

Consider for just a moment that nearly all of the natural sciences build on a foundation supplied by geology, and that geology is based on the existence of magma. If the foundation for nearly all natural sciences rests on *speculative theory*, you would hope that the scientists would consider every possibility, posing important fundamental questions, even those that challenge their favorite theory.

What if there is no magma?

During the late 1700's, James Hutton, a man known as 'the father of geology' proposed the idea that granite, the basement rock underlying the continents, found its origin in a hot molten material we know today as magma. With further research, you will find that Hutton's 'evidence' for this theory was simply his own explanation; he had no direct proof of his assumption. Hutton is one among many influential individuals who impeded real scientific discovery because their new ideas lacked the crucial experiments and observations necessary in true science. As a result, Hutton contributed a weak foundation to modern science upon which the Earth's entire geology would later be built.

Modern science has so widely accepted the existence of magma that scientists cannot even consider the thought of its nonexistence. New evidence in the UM not only proves the nonexistence of magma, but also explains how the removal of this theory dramatically changes many fundamental aspects of all natural sciences.

It might be hard to imagine the far reaching effects caused by something as simple as erasing the existence of magma. Such a change would lead to a new explanation about how rocks form, how we explain the Earth's energy field, and even how we calculate the age of the Earth. This is exactly what the Universal Model does, starting in chapter 5. The UM demonstrates, with actual, observable evidence, that there is no magma.

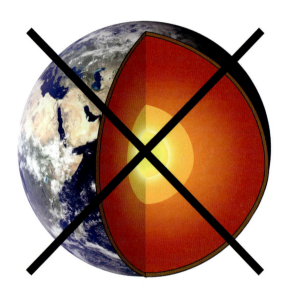

With the removal of magma, UM researchers opened their minds to a whole new list of amazing fundamental questions that led them to new discoveries and amazing new natural laws. Two of these questions were:

If not from magma, what creates lava, and where does the heat in the Earth's crust originate?

The UM provides the true definitions for lava and magma, and uncovers the tendency for modern science to misuse and combine these terms, creating much confusion in geology. The Universal Model covers dozens of evidences of lava's true origin and the source of the heat in the Earth's crust. This physical evidence led UM researchers to two new fundamental natural laws, described in the Magma Pseudotheory chapter.

That chapter also covers other false pseudotheories involving the Earth's origin through accretion, the Earth's radioactive iron core, the Earth's magnetic field, the Earth's continental uplift over time, and many other theories that prove false with the dismissal of the magma theory.

No doubt, the magma theory will continue as a "hotly" debated topic in the coming years because so many other false ideas rest on this one false theory. Once we establish the truth about magma and geologists no longer hold firm to this false foundation, we can move forward with a new, true geological understanding.

The Rock Cycle Pseudotheory - UM Chapter 6

During one summer vacation, our family was touring the American Southwest and stopped at a rock shop. The owner had been a professor of geology at a local university earlier in his life. He was very informative, commented on a number of the rocks on display, and discussed the details about how he believed they were formed. My father asked him why he left teaching. He declared that "Geologists do not know anything about rocks and what they think they know is all just theoretical." His comments were surprising, but his claim about modern geology was astonishingly accurate.

Magma is the supposed material of origin for most rocks, and since magma does not exist, the theory science uses to describe the origin of rocks and the cycle through which they pass must also prove faulty. This means that the modern geology Rock Cycle Theory, which supposes that most rocks are of igneous or a melted origin, is false. Not only was the Earth itself never molten, but the majority of the rocks and minerals we find in Nature were also not formed from a melt.

This Rock Cycle Theory is taught as fact in schools all over the world, earning it a pseudotheory label. Because we understand the truth about magma, we can ask a fundamental question about rock origins.

Can the Rock Cycle Theory be correct if magma does not exist?

With magma no longer a solid foundation, the rock cycle theory is also unsound. In chapter 6 of the Universal Model you will investigate how modern science claims rocks form, and you will also learn that most rocks do not come from the erosional processes suggested by the Rock Cycle Theory. Of course, erosion does occur, but there is much more to the story than what modern geology currently understands!

Faulty foundations have kept us from true wisdom and knowledge about geology. For example, without the wisdom of the true reason why oceans are salty, or the knowledge of where the salt comes from, science can never really understand the crust of the Earth, its landforms, or their origins. Knowing *why* the ocean is salty *is* that important.

The Rock Cycle Pseudotheory

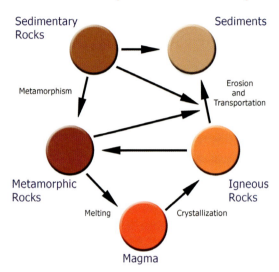

To understand nature clearly and to even begin to comprehend the origin of our planet, we must be able to understand how the rocks and minerals that make up our Earth's crust form. Each rock or mineral exhibits very different characteristics *because of the way each formed in Nature.* Only after exploring how they form and in what conditions they grow can we learn *where and when those conditions actually existed on the Earth*, and why the Earth could not have formed from a continental melt. The next fundamental question that UM researchers asked was:

If most rocks on the Earth today did not form from magma how were they formed?

Chapter 7, the Hydroplanet Model answers this question with simple, observable evidence and explanations, including many new and exciting natural laws.

The Hydroplanet Model - UM Chapter 7

Toward the end of 2005, a think-tank of senior scientists was asked to pinpoint the most critical development in space over the previous 25 years. The conclusion they reached was *the confirmation of large quantities of water ice on Mars*. Water is essential to life and therefore, it is of little surprise that the discovery of water in other parts of our Universe achieved recognition as the most critical development, especially considering the hope researchers have of discovering possible Earth-like habitation or other Life in our Universe. Recognizing water as the precursor to life, scientists have been driven in their search for water with a biological mindset. The Hydroplanet Model chapter demonstrates how water plays host to more than modern science has ever imagined.

The abundance of water throughout our Universe is astounding! The discovery of water on Mars is only the tip of the iceberg when it comes to water discovery in our Universe throughout the last few decades. In 1995, after frozen water vapor kept blurring the camera lenses in a space telescope a million miles from Earth, modern science had to admit that space was not as empty and dry a frontier as they once had thought. In 1998, scientists discovered water ice on our Moon; possibly millions of metric tons of water. Scientists recently admitted to water ice being the most abundant substance found in nearly every comet and everywhere they have searched, even just floating around in space.

As you will see, it became hard for astronomers to look anywhere without finding water. What makes this more astonishing is that we were not supposed to find such an abundance of water per modern science's Big Bang Theory. Hydrogen, not water, is purported to be the most abundant substance. However, the reality of the subject is before us.

Considering the abundance of water in space, the typical person would think that water would have become a central topic of chemistry in star or planet formation. Unfortunately, this is not the case. What is shown, in great detail in the UM, is that star and planet formation, based on the revered Big Bang hydrogen abundance theories, ***does not work***. The existence of water in this quantity directly contradicts all existing Big Bang chemical and star forming theories. This poses a new fundamental question:

> If the Earth, and other celestial bodies were not formed in the Big Bang, how were they formed?

The Hydroplanet Model answers this question with four basic, yet life altering, new natural laws. These new natural laws are paramount to our understanding of how our Earth was formed. The primary reason science has been looking for water outside our Earth was to find possibilities of Life, not to learn how rocks or planets were formed.

The new natural laws identified in this chapter are the direct result of knowledge gained by dislodging the Magmaplanet Pseudotheory from its lofty position and replacing it with the Hydroplanet Model, with its true understanding of the Earth's inner liquid.

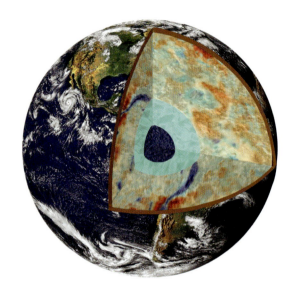

In chapter 7 we examine amazing evidences for the new Hydroplanet Model. Understanding it will enable science to move forward toward a new formation model that answers many mysteries surrounding today's modern geology.

A good analogy to describe the importance of understanding **why** water plays such an important role in the origin of rocks can be likened to the germs on a doctor's hands. Until medical doctors came to understand the role germs played in the health of the human body, they could not advance in medical wisdom. The same holds true in geology. Until we come to the realization of the role water plays in the origin of rocks, we cannot advance in geological wisdom. This fundamental question comes next:

How do rocks grow out of water?

To define the setting where rocks and minerals grow, the Universal Model uses the new term **hypretherm**, defined as the physical environment in which mineral crystals grow, which requires water, pressure, and heat. Thus hydro-pressure-thermal forms the hypretherm. To understand the origin of rocks, minerals, and crystals on the Earth requires knowledge of the hyprethermal environment, which includes five primary ingredients. As with any recipe, if the right ingredients are not present in the right quantity or magnitude, the results will vary. Learn the correct recipe and follow the instructions precisely and you can easily reproduce your favorite cake or crystal.

By simulating Nature and making their own man-made hypretherm, UM researchers were able to grow their own quartz mineral crystals in the same way that Nature does, in a hypretherm. The lab-grown crystals experienced rapid growth, approximately doubling in size in *one day*, not over billions, millions or even thousands of years as modern science insists is the case.

Because quartz minerals represent the most common type of mineral found on the Earth's continental surface, UM researchers understood that these hyprethermal conditions had to be present on a global scale during the Earth's history to produce the variety and quantity of quartz based rocks present all over the surface of our planet.

The Universal Flood - UM Chapter 8

With the knowledge about the necessity of hyprethermal conditions that must have been present on a global scale sometime during the Earth's history, UM researchers were prepared to discover the evidence for one of the most physically documented catastrophic events in our Earth's history. What they discovered was that this event was also the greatest creative event since the Earth's formation, giving us many natural resources such as salt, oil, and precious metals that we use and depend on today. This event is soundly rejected within the world's centers of education because they are distracted by the Magma and Rock Cycle Pseudotheories which requires that we answer this very important fundamental question:

> If this large cataclysmic event actually happened,
> where is the physical evidence left behind to prove it?

This is precisely what the Universal Model shows throughout the Universal Flood chapter. Physical evidence that you can touch, see, and experiment with that proves unequivocally this cataclysmic event actually happened. There is even evidence left behind that tells us *when* it happened. Geologic clues, or **marks** of the Flood, provide evidence that our entire Earth was at one time covered in water. *We find these marks literally all over the world.* The quantity and quality of physical evidence demonstrating a total earthly inundation of water is truly astounding.

Also in chapter 8, you will even read empirical evidence to answer finally the question many children, both young and old beg to know:

What happened to the dinosaurs?!

Clearly, dinosaurs died and their fossil remains almost always lie in some form of flood sediment, but the details of how and why are found only in the UM. Most importantly, correct answers *are* available that make sense, backed by testable scientific evidence, not just theories and flashy Hollywood guesses.

To validate its historical nature and place in time, the physical marks of the Universal Flood allow scientific testing to establish and prove that the event actually happened. For the most part, the scientific tests and observations performed in the UM are readily accessible and simple enough for almost anyone to repeat.

The marks spoken of also provide the evidence that the Flood was a one-time, global event, verified by the observation that the physical marks occurred together, suddenly, during a single point in time. The UM shows that the processes associated with many of the Flood-marks lie dormant today. These marks include but are not limited to, the sand mark, the hydrofountain mark, the erosion mark, the oil and gas mark, the depth mark, and many others, totaling 13 together. There are *many* more marks of the Universal Flood on the Earth today, but these 13 provide sufficient evidence to make the point that the Universal Flood was an actual event.

These marks provide the empirical evidence demonstrating that hyprethermal Flood conditions existed on the Earth's surface worldwide only a short time ago. Rain could never account for enough water depth to create the pressure necessary for the hyprethermal environment.

Where did all the water come from?

Amazingly, the source has been right under our feet all along, just waiting to answer this fundamental question. The Universal Flood was the most crucial geological incident ever to occur on the Earth since its initial formation, and it would never have happened if the Earth was not a Hydroplanet. The chief obstacle barring a scientific explanation of a global flood has been the question concerning the origin of the water. With the new discovery that the Earth is a hydroplanet, we have, for the first time, a mechanism explaining how water covered the entire Earth.

Chapter 8 illustrates an in-depth, eight step mechanism outlining the process of how the Earth was globally covered with water. The physical evidence left behind enables us to see with certainty the scientific reality of a Universal Flood.

The Weather Model - UM Chapter 9

It has happened to the best of us. It's why you decided on an outdoor wedding and the reason you packed your suitcase a certain way for a trip. It's why little Johnny's birthday was set for a pool party and the reason you took off work to drive to your favorite ski resort. You checked the weather forecast, and made your plans. Then the *unthinkable* happens. It rains cats and dogs on your wedding day and during little Johnny's pool party. It snows on your vacation when you packed only shorts and the ski resort never actually got that snow dump you took off work for. It seems that at some point we all ask ourselves this fundamental question:

With major technological advances and years of study, why is the weather still such a mystery and so unpredictable?

Probably for as long as humans have been around, weather has proven mysterious, a plan-wrecking enigma that continues today with modern meteorologists acknowledging that their forecasting is still an inexact science. There is still a lot modern science needs to learn about weather, including the origin of weather itself.

Until scientists can predict where and when major storms and weather patterns will occur, we cannot claim we understand the natural laws driving the weather. Furthermore, we cannot make accurate predictions until we understand *where weather comes from.* As of now, meteorologists understand only that weather is brought on by high and low pressure systems, but they do not comprehend from where the different pressures originate.

Of course, meteorologists point out that their predictions have improved over the past decades, but the reason for this improvement is simple, they can literally ***see*** the weather coming from farther away. Advances in technology, especially satellites, and ever-present commercial air traffic, give meteorologists a constant stream of data and continuous images of global weather phenomena. Obtaining more data via new technology only increases knowledge of weather patterns *as they happen*, but gives little help in predictions made beyond a few days because the ***origin of weather is still not understood.***

Without the dark and dreary clouds of pseudotheories blocking our view, the UM has allowed a geological light to dawn on an entirely new, and more complete understanding of our Earth's weather. In the Weather Model chapter you will find six new natural laws, each based on the Universal Scientific Method, including new experiments and observations. You will read exciting new discoveries explaining the true origin of planetary weather that will answer fundamental questions such as:

What mechanism causes the day-to-day changes in the weather and what is the origin of the Earth's high and low pressure systems?

After reading, many are surprised to find that the foundation of new geology in the UM is also the foundation of a new Weather Model. New evidence shows that the Earth's crustal movement has a greater effect on the weather than previously understood.

Along with new natural laws regarding planetary weather, the true origin of the Earth's energy field is revealed. The false theories concerning a magnetic field originating from the Earth's radioactively heated magma and iron core give way to a new Earth's energy field model. Many fundamental questions for this incredible new claim open a dialog and discuss the physical evidence leading to their answers. Some of these questions include:

With the dismissal of magma and its iron core, what causes the Earth's energy field and why does it exist?

Chapter 9 in the UM also addresses some of the misconceptions about Nature's weather cycles and whether the claims made by modern science, political influence, and media hype are warranted and based on true science. Knowing the scientific facts about whether humans and our actions, or lack thereof, influence global weather is extremely important. The Universal Model answers fundamental questions like these and many more:

What drives weather cycles and why are they never quite the same?

The primary purpose of the new Weather Model is to enable science to reach beyond a mere understanding of the Earth's energy field and day-to-day weather patterns. With a complete Weather Model, meteorology can move towards **prediction** of those fields and patterns. Certainly the Weather Model has great potential in helping us observe how, when, where and why weather changes occur. The beauty of weather is magnified when we understand the Weather Model, and the beauty of Nature is only surpassed by comprehending how it works.

After completing the Earth System, you will likely see our planet in a completely different light. Simple experiences, such as taking a walk to enjoy the great outdoors, will have new meaning. Taking the time to observe and experience Nature with the new UM concepts in mind will allow you to see firsthand the amazing clues the Earth has left for you to find. The new geology model presented in the Earth System is built upon observable evidence and not imaginary pseudotheories. As a result, it is easier to grasp and exciting to finally gain a more complete understanding of our Earth and the many geological mysteries that surround us all.

NATURE IS TOO BEAUTIFUL

NOT TO UNDERSTAND

Chapter 5
The Living System
A True Understanding of all Living Things

"There is more to Life than just science. This is the greatest discovery any scientist will ever make."

Universal Model, Subchapter 13.6

Just as the Earth System chapters helped you to understand the history and origin of our Earth, the Living System chapters will give you scientific evidence into the history of humanity and our origin. The Living System focuses on allowing you to better understand Life with real-time observations.

While reading the Living System you will discover and gain comprehension about the important role history plays with science. Although an integral part of science, the historical narrative advocated by modern science intentionally dismisses much of the physical evidence of our Earth's true history. This in turn leaves us with many holes in our own origin and history, leaving many fundamental questions unanswered. Learning the answers to these questions reveals the basic aspects of Life, which many science experts still do not understand.

Asking fundamental questions and gaining a true understanding of our history and origin will help answer the questions of what, where, and when Life on Earth happened and how we got where we are today. Questions such as: What is the true age of the Earth? Where do fossils come from? When and where did the first human beings populate the world? The answers to these questions help explain the processes and history of Life, giving you the opportunity to comprehend how and why they happen.

By considering current scientific data and by introducing new scientific evidence, the Living System develops a new point of view within which to see one's own origin. Knowing the true answers to Life's basic questions will empower you to detect the errors inherent in current theory, revealing the new purpose as to why Nature exhibits an intuitive order in all things.

The Age Model - UM Chapter 10

Geologists and astronomers love to use the time frame of *billions*; rocks are billions of years old, the Earth is 4.5 billion years old, the Universe is 13 plus billion years old, the number is used so often that we have lost touch with

37

just how big a billion really is. Imagine counting to one billion. One-two-three... It goes pretty quick with small numbers, but once you get to the big numbers, such as 496,342,898, it takes an average of 4 seconds to say the number, translating to 126 years at 24 hours a day without any sleep to get to just one billion!

Or try comparing 1 million and 1 billion. Imagine spending one million dollars at 1 dollar per second. 1 million seconds passes in just over 11 ½ days. Now imagine spending 1 billion dollars at that same rate. It would take almost *32 years*, without a second of rest to dispense of 1 billion dollars! Any age measured in billions is too large for human beings to comprehend, or for the researchers to calculate scientifically with any degree of accuracy.

Throughout our lives, we have been exposed to many theories concerning the age of our Earth and the age of things that lived on the Earth. Most tend to ignore the details of the methods used to determine those ages. Too often, people either accept or reject age theories without understanding or comprehension, partly because modern science has made the details too complicated and out of reach for the average person.

This UM chapter takes details from many books and journal articles related to dating methods and organizes them in such a way that the ideas can be easily grasped. Throughout this chapter, the details of different dating methods and the scientific evidence to evaluate them show which dating methods work and which methods fail.

Today the vast majority of scientists and researchers have never performed dating experiments themselves and are likely unfamiliar with the details published in the UM. They merely accept and rely on the theories propagated in modern science as fact.

As we correct old, unverifiable dating methods, and replace them with new demonstrable techniques, dating can provide consistent and accurate dates which science can use to explain things that were once unexplainable. We must replace modern science's current dating system plagued with fatal flaws and faulty dates with verifiable dates before we can ever hope to understand **when** natural events actually occurred.

The Age Model provides such testable evidences from an Earth chronology vastly different from the one modern science has perpetuated over the last 100 years. For the most part, true dating methods are relatively easy to understand and perform, and they include the most important aspect in dating - *the ability to reproduce accurate age estimates.*

The Magma Pseudotheory chapter presented an abundance of physical evidence that contradicts the magmaplanet theory and after reading that chapter, perhaps you wonder:

Why does science continue to promote the magma pseudotheory?

One of the primary reasons magma remains so firmly entrenched in the theoretical framework of modern science is that *the existence of magma is their foundation for dating the Earth.*

The notion of millions or billions of years that you have heard throughout your life is based almost entirely on the existence of magma. Here are the steps that will be the eventual outcome of the current magma-based dating theory.

1. The Earth's 4.5 billion year age *is based on* the radiometric dating of rocks that supposedly came from a molten melt.
2. The dating of these rocks *is based on* the existence of magma because the dates are started when rocks were supposedly first melted.
3. There is no physical scientific proof that magma exists, and as the Magma Pseudotheory chapter demonstrates, there *is no magma in the Earth.*

4. Therefore, the dating of rocks *is based on a false premise* giving modern science a false date for the age of the Earth.

Magma is the foundation on which modern science has placed its dating framework. A scientific dating revolution is inevitable and when it happens, it will have a domino effect throughout all of science. No field will be left untouched because dating is a fundamental aspect of all scientific disciplines.

There may be those who choose to roll their eyes at this new information and with exasperated sighs, they might ask this question:

Is the actual age of the Earth really that important?

The answer is a loud resounding *YES*. The dating lens through which you look determines how you view not only the Universe but ultimately how you view yourself. With that in mind, it becomes critical to discover and confirm the *true dates* of all things. Otherwise your perception of the world around you, and more notably, your perception of who you are, will be flawed.

The Fossil Model - UM Chapter 11

Fossils are a fascinating scientific subject enjoyed by both children and adults alike. They are touchable, making it possible to experience them firsthand. Science can only partially describe what fossils are because they have never correctly understood how fossils form. A fossil is a rock, but its definition cannot be complete if we do not understand how the rock was made. The Fossil Model chapter includes the correct definition of a fossil, which includes **how** the fossil formed. The purpose of the Fossil Model is to demonstrate the origin of fossils and enhance our understanding of the remains of once-living organisms that have now turned to stone.

For hundreds of years, scientists have relied upon the illusion that they could never *form* a fossil. This is because they believe fossils take millions of years to form. Because of this, researchers have put little effort into attempts to produce fossils. They teach and believe that fossils are formed when organisms die and sink to the bottom of streams, oceans or swamps and then over millions of years are slowly turned into stone. This theory drove UM researchers to ask this fundamental question:

Where are the **in-process** fossils that are forming in the streams, oceans or swamps **today**?

Most people can answer the question, "What happens to living organisms when they die?" They break down: bones disintegrate, flesh and wood decay. Dead matter is fairly quickly broken down, especially in a water environment.

The UM demonstrates that there are no large quantities of active, in-process fossils, because almost all of the fossils we find *only* on the surface of the Earth today were all made at one time, in a single, worldwide catastrophic event. *Not* over millions of years after things had died and sunk to the bottom of bodies of water as modern science teaches.

Paleontologists are unable to grasp the origin of fossils because they do not know the true origin of rocks. Petrified wood, the most abundant fossil on Earth, consists primarily of the silicate mineral quartz. After UM researchers discovered how to form quartz, attempting an experiment to create their own petrified wood was the next important step in identifying the conditions in which petrified wood formed in Nature. Once we determine exactly how minerals *are* made, we can know with certainty how they *were* made in the past.

Over a four-year period and after 70 plus attempts at getting the fossil recipe exactly right, UM researchers produced the *first* known man-made sample of fossilized wood. Remarkably, the fossilization process, from start to finish, took little more than *two days, not millions of years*! Because modern science depends almost solely on unproven theories involving millions of years, it believes it impossible to reproduce actual fossils in real time. Conversely, the UM employs hands-on testing, observation and evaluation, and by questioning the long-held assumptions, was able to produce man-made fossils, using repeatable processes anyone who wants to, can do.

For over two centuries, paleontologists passed down the same fabricated story; that fossils appear in layers of rock that succeed one after another in an evolutionary order; the lesser advanced species in the bottom older layers, the most advanced species in the younger top layers. It is a common sight in many natural history museums to see a reconstructed fossil record showing the fossils in their corresponding time periods dated back millions of years in conveniently arranged layers. These beautifully crafted artificial exhibits never answer this fundamental question:

Where in **Nature** do these layered fossils show the depicted evolutionary timeline?

Fossil Record
Timeline Pseudotheory

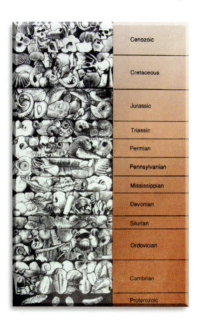

My father once asked several geologists and paleontologists where he might go to see the actual layers of successive fossils in a natural setting. None could give him a location, and as it turns out, the *only* place anyone has ever seen an evolutionary ordered series of fossils is in the artificial museum exhibitions where they 'create' whatever layer they desire, including whatever fossils they want to show. Only in the artificial layers can anyone see fossils ordered according to evolutionary theory; they *exist nowhere in the natural world!* Real science *demands* that the represented layers be documented somewhere in Nature, if they are an actual representation of the natural fossil record, otherwise, it represents a fraudulent modern science claim!

Modern scientists have mastered the art of 'talking about' their theories but not demonstrating them in Nature or in the laboratory; just theoretical talk with effectively no prediction and no testable or observable experimentation. In chapter 11 in the UM, the actual scientific evidence showing how fossils form includes more than merely talk and fancy exhibits professing how we 'think' fossils were made. Instead, we see the fossilization process *demonstrated* by actual, repeatable experimentation.

The Evolution Pseudotheory - UM Chapter 12

The theory of 'millions or billions of years' refers to what scientists think of as **deep time**. This modern science invention helps justify many other favorite modern science theories, especially the Theory of Evolution. In today's modern science circles, evolution represents solid fact by most scientists, and they claim that nothing in biology makes sense without it. For them, its principles are so vital that there is no point in considering any other alternative.

When a theory such as evolution becomes so widely taught as hard fact, fundamental questions that challenge it are routinely and flatly rejected, without even a second thought. So we must ask:

> What physical, scientific evidence supports
> the theory of macroevolution, or speciation?

In this book, and in the UM, the general use of the term 'evolution' references the unproven idea of macroevolution or speciation, not microevolution or adaptation, which is demonstrable.

We must not allow the calculated mingling of **microevolution** truth (also known as adaptation) and **macroevolution** error (also known as speciation) to confuse the world any longer. Microevolution describes changes *within* a species, and it represents factual, observable, and repeatable evidences and processes, such as the observed changes in various breeds of dogs. Conversely, macroevolution, where a living organism becomes an entirely different species, has *never* been observed by anyone, anywhere. It is not repeatable, and thus, ***scientifically, it is not factual***.

The pages of the Evolution Pseudotheory chapter include evidence not previously discussed in modern science circles verifying the falseness of the theory. Based on discoveries in the Fossil Model we know exactly how fossils form, and we know the how long the process takes. Fossilization in just days destroys the assumption that fossils take millions of years to form, positively repudiating modern sciences' claims and removing an integral defense scientists rely on to validate evolution.

Evolutionary theories, like the dating theories, rely heavily on assumptions. Perhaps the greatest assumption of all is that the theoretical processes heralded by modern geology and modern paleontology are correct, despite all of the evidence to the contrary.

Within these theories, we find a corresponding fatal flaw – *time*. Without sufficient time, discussions about *how* evolution might happen become obsolete. Without the deep time of billions or millions of years, *evolution could not exist*.

Evolution has never been a 'science' where laws and experiments demonstrate evolutionary events. Evolution, a philosophical belief, came about in part because there was no other scientific explanation of how living things became rocks or fossils, leading scientists to *suppose* fossils supported their *unproven theory*. With the discoveries and natural laws introduced in the UM's Earth and Living Systems, and with the ability to test and observe actual fossilization processes, we can lay aside the theory of evolution in favor of truth. For only true science will ever displace false theory, including evolution.

As a theory, evolution must stand the scrutiny of scientific fact. We should leave out personal opinions and biases because ***they are not science***. Despite technological advances, countless anomalies befuddle researchers, and Nature stubbornly refuses to reveal answers to questions when those questions are built upon false premises.

The interdependent fields of biology, geology, paleontology, archaeology, and a host of other disciplines experience the effects of shared reality or falsity, each field having a profound effect on the others. Removing the Magma Pseudotheory from geology affects drastic changes in almost every field in science. Dismissing the Evolution Pseudotheory will have a similar effect, perhaps even greater, as it is directly connected to how we ultimately view our purpose, life's origin, and ourselves.

The Living Model - UM Chapter 13

The Living Model chapter is a scientific exploration of Life and its symbiotic interactions. Intended to replace modern science's evolution theory, the Living Model successfully explains real-time observations of the living world that have stood the test of time. Discover four new natural laws pertaining to life while taking an in depth journey into the exciting new Microbe Model, learning the simplicity of the true origin of all humanity.

Life's inherent purpose, coded in endless interconnected organized patterns and relationships, is evident in all living things. This sparks a fundamental question long asked and often pondered by curious children and reasoning adults alike:

From where did Life come?

If not from evolution, then from where? Though multitudes ponder and debate endlessly, knowing the correct answer is of extraordinary consequence from a scientific perspective. Portrayed as mere accident, Big Bang and Evolution theorists suppose Life is too messy to understand and its origin impossible to prove in real time because their theoretical processes take millions, even billions of years to actualize. However, true science demands observation and these excuses only leave us with an uncertain view of our own life and purpose as well as many unanswered questions on the origin of Life.

Is Life really that messy? Perhaps it is if one does not have the answers, and answers are hard to come by in a science void of observation. But in Nature, the simple truth is that truth is simple. In contrast to Evolution's uncertain view on Life, the Living Model presents simple, natural laws that illuminate how living organisms work and interact in Nature.

The Living Model introduces amazing new truths about the most abundant variety of species on Earth, *the microbe*. You will discover the significant role that microbes play in both the formative processes on our Earth and in our own bodies.

Microbial research is one of the most promising areas of new science and medical study. New findings over the last few years from modern medical science, along with the UM's new discoveries concerning microbes, will influence the daily health of human beings in many ways. As such, stated knowledge about essential microorganisms that comprise as much as 90% of the cells in the human body (by number) becomes very important. It is an exciting time in the budding field of microbiology, as it stands on the threshold of incredible discovery, potentially greater than in any other field.

We define biology as the study of life and living organisms. In an attempt to organize such a broad range of organisms, they are commonly broken down into different species. Biologists, beginning with Charles Darwin himself, often claim ignorance when asked to define the term species. Perhaps the reason they struggle is because the true definition of a species, given in chapter 13 in the UM, defines a barrier that a species cannot cross, in essence the true definition of a species is anti-macroevolution.

Human beings do not create natural laws - we only discover them. Order does not come spontaneously from disorder. The UM acknowledges that a *full knowledge* about the origin of Life and the natural laws that govern the physical world remain undiscovered in modern science simply because *some things are beyond science*. However, it is the responsibility of human beings never to stop learning and discovering new truths in the hope of a better understanding of the world around us, including ourselves.

The World History Model - UM Chapter 14

Humans record and learn from their past and have done so for thousands of years. We do it in words, written or engraved, and through oral tradition. In addition to the records left to us by our predecessors, the Earth also has left a physical record for us to observe. Many scientists invest entire lives striving to understand past civilizations and human origins by studying these records, because *knowing is important*.

Although it seems an unlikely chapter in a science book, the purpose of the World History Model is to honor and reexamine history within a scientific mindset and to ask the fundamental questions:

What is the relationship between history and science? Is history important to science, and if so, how important is it?

In the World History Model chapter, you will evaluate evidence of this relationship in new ways as you set out to answer these questions and others. Chapter 14 dives deeper into the meaning of Scientific Truth, which we define as the knowledge of what is, **was**, and so far as is known, what will be. Remove the word "was" from the definition and you no longer have scientific truth. Indeed, without history, there is *no* true science because true science requires the *test of time*, or historical data to test and evaluate observations.

Real history is a record of *true* events. You can apply the Universal Scientific Method to the investigation of past events to establish a true historical record. Humans have recorded history from the beginning of time, a subject discussed in the World History Model chapter. Of course, the record of events can be clouded according to the experiences and perspectives of the recorder, and it is sometimes done without respect to the truth.

The only way to establish the truth of the event's reality, or the difference between history and myth, is to examine the observed record. As you do so you will find physical evidence from many sources, each with its own part of the story. Historians continue to update the exact details of how or what happened during the event as they are engaged constantly in historical improvement based on current facts and the discovery of new knowledge.

Many aspects of our Earth's history have been under attack for centuries by revisionists and it will likely continue. Many scientists will simply reject the physical records made available throughout the World History Model, and throughout the entire UM. Even with *extensive evidence*, there will always be those who choose to remain inflexible, but it is time for scientists to respect true history as science.

The World History model references many recorded histories, however these records are by no means a complete historical account of any group of people, including those spoken of in the record, and especially not the whole world. A common misconception among cultural groups is that their history is at the center, above all other histories in the world. The reality, of course, is that history occurs everywhere at once. Some records are more complete than others. Sometimes the records kept by one group of people contain information and details of surrounding groups, even better than the surrounding groups kept themselves.

In this chapter, you will read about evidence for some of the greatest scientific events that happened in our Earth's history. The greatest events are those that have had the most lasting influence on the whole of humanity. These events left a physical historical record of which the World History Model explores in depth, including eight historical records: the Hydroplanet Record, Universal Flood Record, Continental Drift Record, Bottleneck Record, Family History Record, Language Record, Archaeological Record, and the Scientific Discovery Record. The goal of chapter 14 is to demonstrate scientifically three major world history events important to every human being through the examination of these physical, historical records.

The World History Model is both a new and an old model. New in that it documents for the first time, mankind and the Earth's history from a comprehensive set of scientific records. Old in that no one can change history or the events that happened in the past. You can choose to understand them, to study them, or to reject them, but you cannot *change* them or the fact that they happened.

The Clovis Model - UM Chapter 15

Looking back across hundreds of extraordinary UM scientific discoveries, the very *first* UM discovery was that of the Clovis Model. The official beginning of the UM came in 1990 when our family had been traveling through northern Arizona. My father walked into an antique store and spied a beautiful display of arrowhead points behind the counter. He asked the storeowner how much she wanted for the display and the impressive points it contained, but was met with a laugh as the owner informed him they were her personal finds and not for sale.

From a very young age my father remembered his own father sharing two arrowhead points he had collected as a child. The feeling he felt while holding those points in his hand never left as he thought about the ancient person who had made them and what their use was. He had always dreamed of finding his own point but not knowing where to look, he never had.

When the storeowner saw his disappointment, she took him to the back of the store and gave him an old jug full of broken arrowhead pieces to sort through, allowing him to buy as many as he wanted for fifty cents apiece.

The experience renewed his fascination with ancient artifacts and it became the catalyst for many fundamental questions, each leading him on his UM journey:

> How old were they? Who made them and where did the people live? What were they used for? What kind of material was used and how did they make the artifact?

The questions flowed and they would not stop until he found the answers. My father learned that to truly understand his archeological finds he required an in-depth knowledge about how the tools were made. Thus, he set out to learn how to make these types of artifacts and that led to questions about the type of rocks used to make the artifacts, which in turn led to a desire to know more about how those rocks and minerals formed.

It was hard to study geology and ancient artifacts without finding fossils increasingly fascinating, since fossils seemed directly associated with many of the finds. The fascination with fossils led him to ask questions about how they formed, and thus began a domino effect of scientific discovery brought on by asking questions and objective research, culminating with the comprehensive Universal Model project.

His own story provides for us an example of how a small and simple event can inspire great changes in scientific discovery. Looking back over the years, hundreds of these seemingly insignificant experiences led to scientific discoveries never before seen, heard of, or even imagined.

These discoveries remained unknown because modern researchers did not know the true history of the Earth's past, due in large part to three pseudotheories, which cloud the minds of modern scientists. The Clovis Model addresses the three pseudotheories, which are the Dating Pseudotheory, the Ice Age Pseudotheory, and the Evolution Pseudotheory and their hold on the scientific world, which has led to much confusion and contention related to fundamental questions such as:

Who were the first Americans?
Where did they come from?
How did they get there?

Chapter 15 answers these questions and discusses the first 1600 years of human existence. Geographically it focuses on the North American continent, explaining the First World Model as the place where humans began. With new scientific research supporting many extraordinary claims, it defines the Clovis people and their culture, expounding on the five reasons why Clovis points are unique and unlike any other artifacts produced by ancient humans.

The uniqueness of the Clovis culture and the special place it holds in our common human history is outlined and explained using research from a number of scholars from around the world, coupled with new UM observations. By combining the knowledge of the Universal Flood's physical marks and the wisdom acquired by forming the first man-made fossils, UM researchers were able to answer many more fundamental questions because they had discarded false theories when new information came to light.

The Human Model - UM Chapter 16

The last chapter in the Living System represents the pinnacle of the UM's big picture, the human being. As you follow the inherent order of nature, which leads to the purpose of humanity, you find yourself a part of something much bigger than any one of us. The Human Model is the capstone of the Living System and as such it answers some of the most important fundamental questions such as:

What does it mean to be human? What makes a human being different from all other beings?

In the Human Model you will learn why we describe human beings as the highest order of intelligent life on Earth. Intelligence is what sets us apart from all other beings. The mere fact that you are reading this book is evidence of this. Still, how do you truly define a human being?

In modern science today, to be human means that you came from nothing. From nothing came the Big Bang, which accidently caused chemicals to combine into some type of bacterial form, which after a few million years mutated into a fish-like creature that left its watery home, until a few million more years when it evolved into apelike ancestors who eventually evolved into us. This is how modern science defined humanity; mutated organisms that came from nothing. Most people find it astonishing that modern science still

believes this, but they do. And they refuse to accept the alternative physical scientific evidence that disproves their 'from nothing theory' and substantiates the true order of Nature, our Universe, and our own humanity.

The grand objective as human beings is to broaden our understanding of our origin, our destiny, and the world around us.

For now, and the foreseeable future, the Earth is where human beings call home. Finding the purpose and the link between our Earth and us is important, and it is the source of an important fundamental question:

How are we connected to this Earth?

Learning a truth or gaining wisdom in science is all for naught if we cannot integrate it into our personal life. Without a personal connection to Nature, there is no interest in Life. This is the chapter where we discuss the science of being human.

Prepared with the understanding that humans do not create natural law, we only discover it, the Human Model discusses the topic of natural coincidences versus natural order. This chapter puts the incredible ordered parts of our Earth in perspective. Things such as the size of the Earth and its perfect preciseness; any smaller and we would have insufficient atmospheric pressure to breathe properly; any larger and the pressure would intensify beyond our ability to withstand it.

These physical facts seem unnoticed in the modern science world, as if they had nothing to do with a definite purpose and natural order, deemed only a coincidence that came from nothing more than chance.

In the Human Model, you will read about an in-depth emphasis on human life itself, including discovery in medical and political science. As human beings, we must place priorities on *human life* above all other scientific ventures. For the purpose of Life is the *most important knowledge* we will obtain; our own life's experiences are what give life meaning.

The Living System puts Life in a completely different perspective. Simple answers to some of Life's biggest questions will give new meaning and a new foundation to all fields of science. Gaining a true understanding of your own history and origin will allow you to view Life within a new paradigm. Built upon observable evidence and not imaginary pseudotheories, the Living Model presents the opportunity to grasp and finally gain true answers to Life's basic questions, allowing you to see the inherent order in all things.

THERE IS MORE TO LIFE THAN JUST SCIENCE.
THIS IS THE GREATEST DISCOVERY ANY
SCIENTIST WILL EVER MAKE.

Chapter 6
The Universe System
A True Understanding of Our Universe

"The most powerful scientific discoveries are the simplest."

Universal Model, subchapter 19.4

Comprising the third and final UM system, The Universe System examines some of the most basic concepts of physics and chemistry. Physically speaking, you cannot discuss anything bigger than the Universe, which appears larger every time scientists build a bigger telescope. We call the branch of science that studies matter and energy, Physics. While chemistry studies the properties and reactions of that matter. The Universe System also covers primary concepts in astronomy and cosmology triggering a surprising reevaluation of our night-time sky.

Asking fundamental questions and gaining a true understanding of the Universe helps answer the questions of what, where, and when, such as: What is mass? Where does energy originate? When will the Earth stop spinning? The answers to questions such as these help explain the Universe's processes, allowing us the opportunity to comprehend how and why they happen. The Universe System begins by discussing the three basic units of measurement, moving on to Einstein's Relativity Pseudotheory, and ending with many amazing new Universe Models.

Prepare yourself for fundamental modifications to physics, chemistry, and astronomy, a veritable upheaval in these fields of science is about to happen, forever altering them. With these new discoveries, the way in which we comprehend the Universe and our place in it will never be the same.

Essential Measurement - UM Chapter 17

Humans have always measured. When ancient man traveled, built shelter, or planted crops, he measured. As soon as you start to describe your own experience with Nature, you begin to measure. When you buy gasoline, you measure the volume. If you want a cut of meat at the grocer, is it not weighed? What about how hot or cold something is, how tall or short, how bright or dull, fast or slow? All of these descriptions are based on *measurement*. The more accurate our descriptions of measurement, the more correct our understanding of Nature.

We even measure things that we cannot see with our naked eye, such as galaxies using telescopes and bacteria using microscopes. Inherently calculable, measurement forms the essential part of testing, observation, and evaluation.

You will begin this chapter by learning the fundamentals of measurement and by seeing how science measures things today. You will also learn just how important true and precise measurement is to science. Measurement lies at the foundation of any scientific discussion, *for without measurement there is no science.*

We have come to expect accuracy in measurement in our day-to-day lives. How would you feel if every time you filled your gas tank, the pump registered several more gallons than you actually received? Or what if you were trying to lose weight and went to one doctor whose scale registered 220 pounds and on the same day, another specialist weighed you in at 225 pounds? Both instances would invoke feelings of frustration, deceit and for some, even anger. Why should it be any different in science? This thought led to a very important fundamental question:

What if some of the fundamental measurements in science are not correct?

As you read, you will understand that modern science's knowledge about the three essential units of measurement, weight, length, and time, remains surprisingly incomplete, leaving us with more questions and theories instead of answers and natural laws.

UM researchers are always striving to evaluate open-mindedly and since some errors might seem rather small in the grand scheme of things, this fundamental question needed asking:

Are small variations in measurement **that** important?

Because Nature follows natural law, our measurement of weight, length and time should not only be precise, but also accurate *every time.* This chapter explains that important, precise measurements in weight, length, and time can lead to some of the deepest understandings of the nature of our Universe. Comparatively small incorrect measurements can invalidate massive amounts of important evidence and mislead generations of scientists, as well as the public, who depend on accurate measuring.

The Mass-Weight Model - UM Chapter 18

The Mass-Weight Model sets the stage for the UM to redefine physics. The changes begin here in this chapter, continuing throughout other Universe System chapters. Weight is the first basic and common unit of measurement, however, a correct understanding of mass and weight is not common at all, even among scientists today.

Along with the historical background of weight measurement, this chapter introduces details of multiple experiments performed and observed by UM researchers, that transforms the way we evaluate mass and weight.

The definitions of mass and weight are reanalyzed and corrected. These new definitions place the physics of the UM on a completely different course of understanding as to what actually constitutes mass and weight, and it helps answer this fundamental question:

Why are the correct definitions of mass and weight so important?

A red flag should signal a warning whenever we can no longer depend on simple concepts such as mass or weight, especially if someone tells you to simply "memorize what these words mean." Any single subject should have a simple, single definition if we really know what it is.

In our everyday experience with weight, we judge an object heavy or light depending on how many pounds or kilograms it weighs. In every society, for comparison purposes, they must have a weight standard. What would happen if that standard changed, even by the slightest amount?

The international standard for weight, the kilogram, originally saw definition in 1795 with the manufacture of an actual prototype during 1799. Subsequently, curators kept it secured and under a sealed glass container behind vault doors until in 1948, with the commissioned reweighing of the kilogram. Much to the scientist's surprise, the kilogram no longer weighed one kilogram; it had gained mass. Something scientists say is not supposed to happen. After reading this chapter you will comprehend why and how this

happened, something even modern scientists cannot do.

You will find new UM definitions for both mass and weight that align for the first time to what you can actually observe and quantify with experimentation, along with the exciting discovery of new natural laws.

A corrected density and dramatically lower Earth's Mass, as well as mass adjustments to other celestial bodies in our Solar System, appears in the Mass-Weight Model chapter. The mass of the Earth and the gravitational constant, known as 'G' is inversely proportional, meaning that reducing the mass of the Earth requires an increased gravitational constant. Presented in this chapter is the ongoing investigation needed to uncover certain errors related to the supposed gravitational constant 'G.'

The mass and weight issues describe the compelling need to repair physics. Over time there is hope that a crisis in physics will make the need very apparent for the new, simple and correct Mass-Weight Model as presented in the UM.

The Length Model - UM Chapter 19

Length, the second basic unit of measurement defines the size of an object and societies have used some form of it for thousands of years. The way we determine length continues to change as new technology emerges and as our ability to see farther and smaller improves.

Measurement by royal cubit (the distance from the elbow to the tip of the

ruler's middle fingertip) gave way to devices of better consistency, such as rules and gauge blocks. Modern physics discovered that the speed of light in a vacuum remains extremely precise, and uses the time it takes light to travel a distance of one meter as the international unit of length. Extrapolated to great distances, a light-year describes the distance light travels in one year; and astronomers favor this unit of measurement to refer to the vast expanse of the heavens. However, as you will discover in this chapter, the Earth does not orbit the Sun in a vacuum and in fact, there is no pure vacuum in Nature. UM discoveries related to this fact change dramatically how science must view the accuracy of the 'speed of light' in Nature.

Using light to measure length has long been a part of the human experience. Anciently, the astronomers used star light to measure the diameter of the Earth and today researchers use starlight to measure the Universe. Presently, astronomers, cosmologists, and physicists base Solar System and Universe length measurements on the speed of light 'constant' expressed as 'c' in mathematical equations. However, because there is no true vacuum in space, does that speed change, asked open minded UM researchers. Furthermore, UM researchers asked this fundamental question:

If light speed in Nature is not constant,
would this change our perspective on the length
of everyday things we observe with our eyes?

The answers to these questions come from simple, new experiments and discoveries concerning the light-speed relationship. The new insights will have a profound impact on the optics and physics worlds, and they are destined to replace old theories that have never adequately explained many observational anomalies.

Because we observe length with light, if the properties of light change, so can the visual perception of length. Measurement is inherently comparative; it represents the distance between two objects. Our perception of measurement requires that we compare it with other forms of matter. The comparison of a rock to surrounding known objects helps us to discern the rock's size and its distance from us. What happens to the size of the rock as we move away from it? Of course, it appears smaller, but does that happen in a linear manner? The surprising answer is that it does not, which has astonishing implications for the science of perception and measurement.

Other topics in this chapter include an evaluation of light's basic properties, with reflection and refraction providing ample opportunities to conduct important, paradigm-shifting experiments. The Refraction Index Error discusses the persistent error related to modern science's explanation of how light bends when passing through a medium. New UM experiments show that the old Law of Refraction cannot be correct, and replaces it with a new Law of Refraction Index.

The subject of this chapter spawned many experiments, most of them simple enough that anyone can perform them at home. Each quantitatively confirms the extraordinary claims recorded in the UM that answers questions previously unanswered.

The Time Model - UM Chapter 20

The most frequently used noun in the English language, time, is a fascinating subject to contemplate and one that plays a significant role in everyone's life. How often do you think about what time is? Interest in the study of time waxes and wanes like the seasons. Today it occupies an increasingly misunderstood place in modern physics, primarily because of the dissemination of false theories the experts herald as fact.

Sadly some physicists even believe that fundamentally, time does not exist. You will come to see clearly in this chapter that time is as authentic as life and death. Time does not defy mortality, it defines it. This chapter shows that time is real and observable, and is an essential part of all existence. Because of this we must consider this fundamentally important question:

What is time?

After reading this chapter you will learn that all measurement comes down to time. Time is not an abstract, fictional piece of our imagination as some theoretical physicists would have us think. Every atom in our bodies, and in fact, all matter is based on time.

If you ask any two people what time is, you will most likely get two different answers. When this happens we know that we must properly define the word so that everyone can use the same definition, thereby alleviating confusion. In Webster's Dictionary alone there are 54 definitions for the word time. Time is so common we tend not to even think about it, yet not knowing what time is and where it comes from precludes understanding about science and Nature.

Put simply, modern science does not know how time works because they do not know what time is. By misunderstanding time, three basic concepts about time remained hidden; we discuss these concepts in-depth in the Time Model chapter.

Despite modern science's inability to clearly explain or define what time is, time is not necessarily elusive. It surrounds us every day with concepts easy to grasp; all matter exists tied to Nature's Clock in time throughout the Universe. We cannot measure acceleration without time, and without that, we cannot gain an understanding about energy. Without the knowledge of what constitutes energy, we cannot know matter, and without the knowledge of matter, we cannot understand the Universe. Therefore, time establishes the essential base unit from which to begin our quest to understand the Universe and our place in it.

The UM discloses one key quality of time that science has yet to identify, and once revealed, appears evident throughout the Time Model chapter. Think about a winding watch, or a spinning top, or balancing a spinning basketball on the tip of your finger; they all share something. Stop winding the watch, spinning the top, or tapping the basketball and the motion eventually stops because of friction. Herein lies a Time Model fundamental question:

In Nature, how do spinning objects such as our Earth, keep spinning?

This chapter presents evidence why the Earth's axial spin and its orbit around the Sun should be slowing—and then why it's **not slowing**. All matter requires an external energy source to keep time and the Earth is no different. This chapter reveals the external energy source, including with it explanations and supporting evidence to establish new natural laws about time.

Throughout most of our history, time has been an important component of Nature, but modern physics supposes fundamentally that time doesn't exist because of the Theory of Relativity. Time expresses reality and reality is time, and this will become clear as you read through the Time Model chapter, encouraging you to think about this essential measurement as you never have.

The Relativity Pseudotheory - UM Chapter 21

Describing the Theory of Relativity is not an easy feat, even for scientists who believe it.

This chapter is about Albert Einstein and his famous theory; his attempts to define it and why the theory lacks veracity in science. You will see an abundance of evidence for this extraordinary claim as well as an answer to this fundamental question:

Was Einstein actually a scientist?

Real scientists perform experiments that demonstrate scientific principles or processes in hopes of discovering truths which describe and explain Nature better than previous theories. A philosopher merely poses ideas without accepting the responsibility to prove it with repeatable experiments. No record exists where Einstein performed a single experiment himself. Known for his personal belief that his physics work did not require a laboratory, Einstein supposed his physics required only a logical system of mathematical thought. He did not think he needed to do hands-on experimentation. In this chapter, the UM explains why true physics demands both logical thought and experimentation.

Branded as the icon of modern science and selected as Time Magazine's "The Person of the Century," Einstein is recognized as the most popular pseudo-scientist in the world. However, Einstein was chosen for this prestigious title not for what he did, but for what everyone thinks his theory did for science. While many scientists believe Einstein's theory stands proven, this chapter will show there is no tangible, physical evidence proving the theory. Meanwhile, the public hears the deceptive claim that the Theory of Relativity is

fact, while scientists continue intently working to find convincing evidence, such as black holes, to prove it.

With each successive failed experiment, scientists toss around scientific alibis to explain why this experiment was not successful "but came close," or that inconclusive test "gave us new data to study." Their futile and ever more costly mega-projects bear little, if any, fruit. Meanwhile, the weight of evidence against Relativity grows and the flimsy pillars holding it up show signs of collapse.

In this chapter, you will learn that scientists know that if Einstein falls, so too will the pseudotheories built upon his false ideas. Throughout the chapter, you will read about the deception used over the years to try to convince the public that this theory remains viable, important, and proven.

Although the distortion and hiding of truth did not begin with the Theory of Relativity, the public can learn from the experience to always question new scientific concepts with an open mind. Follow the money trail and try to determine the agenda for yourself; is the intent to discover truth or is it something else.

Finally, the UM reveals conclusive reasons why the Theory of Relativity does not stand on solid ground, listing nine false conceptual pillars that can no longer stand weight of evidence against them. As those old pillars crumble the icon of modern science, Einstein, will not survive the fall.

The Energy-Matter Model - UM Chapter 22

This chapter covers four basic forces in nature. We describe these forces using the three fundamental units of measure; weight, length, and time. This chapter may prove one of the most powerful and exciting chapters in the Universal Model because it typifies the new world of discovery heralded by the UM, capitalizing on the most basic concepts in physics while describing and explaining Nature in new ways anyone can understand and comprehend.

One of the big reasons today's science has no real understanding of what energy actually is stems from the belief in Einstein's theory of relativity. His famous equivalency equation, $E = mc^2$, has everyone supposing that energy is equal to mass times the speed of light squared. As crazy as it might seem, you will learn in the Universe system that mc^2 does not represent anything empirical or real, which is why physicists do not really know what energy is. Moreover, we know mc^2 is incorrect because energy is very real; we observe it every day, and that leaves us with this fundamental question:

What is energy?

We often use the word energy to describe electricity, light, heat, or power, but these represent macro descriptions of *types* of energy, they do not tell us what energy is. Science deals with energy every day and it stands as one of the most fundamental topics in the science classroom today.

$$E \neq mc^2$$

The purpose of the Energy-Matter Model is to describe and explain what force, energy, and matter are. Additionally, it replaces much of quantum mechanics by introducing many new natural laws. This model will also replace Einstein's Relativity Theory with a better description of both energy and matter.

Energy and Matter comprise the most basic concepts in all of science—they define the study of physics. You may find it curious that science really has no concept of what they actually are, which is an amazing and extraordinary claim posed in the Energy-Matter Model chapter. The pages are filled with many quotes from the scientists confirming the truthfulness of this claim. It is only through a correct understanding of energy and matter that we can progress scientifically toward describing and explaining all things.

Why can't modern science define energy and matter when they are such a fundamental part of nature?

You cannot define a thing if you do not know what that thing is, and as of today, there is no universally agreed upon scientific definition for the words matter and energy. In effect, modern physics lives in a mythical world because its foundational beliefs spring from theories and principles that one must imagine. Thoughts inside Einstein's or anyone else's head do not constitute scientific truth, and without truth, there is no reality.

Science represents a tangible pursuit, so for something to exist, the senses must perceive it. This is the only way matter can be observed, through the five physical senses. Atomic matter is attracted to each other when it has resonant energy or motion. When millions of like atoms bond in an orderly form, we can observe the resulting matter with your eyes, and if large enough, with our hands. All of our senses come into play as matter organizes into aromas, tastes, and sounds. The five physical senses give us the ability to experience mortality yet, if you cannot explain their existence, you do not really know what they are.

Perceived matter exists because it is organized. Organized atomic matter moves in an orderly motion. This is why all organized matter is perceived, it has order. Matter becomes perceivable when it is organized into structures that are quantitative and reproducible. This chapter not only establishes that there is order in all things, but that through organized matter you will also find organized energy.

To define energy correctly you must first understand its abundance. To do this, we begin by looking at the smallest particles of matter in which we can measure energy, examining ever-larger instances of matter, all the way to the entire Universe. The answer to defining energy consists of only two words that follow true and simple principles. In Nature, the simple truth is that truth is simple, and the simplest correct explanation is the best.

This chapter provides new understanding about light and its origin. As science progresses in its understanding of some things, the physical properties of light and its accompanying mysteries has only deepened. This is because a correct model accounting for the duality of light being both a particle and wave cannot be explained. Light has fascinated humans for millennia. We take the ability to see for granted, but it truly is a marvelous capability. The cosmos, bathed in constant celestial light; the Earth rolls on its wings basking in the light particles of an ever-shining Sun; we glory in light, yet it remains a mystery—until now. For the first time, the UM explains its physical orientation.

Atoms also remain a mystery in science even though modern scientists would have us believe that they have atoms completely figured out. This chapter redefines the simple atom, showing plenty of evidence that modern science does not know everything it claims to know about atomic matter.

We visit and expound the Law of Gravity in this chapter. We recognize gravity in our Solar System because of Newton's observations and his calculations about the effects gravity has on objects. Despite all of his knowledge, neither Newton nor modern science has been able to explain:

1. *Why* objects are attracted to each other.
2. *Why* gravity is such a weak force.
3. *Why* gravity does not work the same on a galactic scale.
4. And *how* gravity relates to the other forces.

This chapter discusses these fundamental thoughts, presenting more new significant scientific discovery than any previous chapter. Through the UM, modern physics stands to experience complete reform. Perhaps it can migrate from being the author of confusion—to the author of understanding.

The Air-Water Model - UM Chapter 23

Although we cannot see air, we can feel it. You fill up your tires with it and your perfectly finished hairstyle can be blown away by it. We cannot imagine living in Arizona without cold air conditioning; sitting on the beach in Hawaii just wouldn't be the same without a cool breeze. Because breathing and feeling air comes so naturally to us, it is probable that few people actually stop to consider some of the fundamental questions about air, such as:

What is air and where does it originate?

Why would we ask what air is? Long ago, scientists determined that the Earth's atmosphere consists of about 78% nitrogen, 21% oxygen, 1% argon, a variable amount of water vapor and minor amounts of CO_2 and other gases. However, do the percentages of gases really tell us what air is? They do not.

Science supposes that air and other gases are made up of random, fast-moving molecules, but this has never been observed. In fact, new UM experiments with air and water help us understand and comprehend these fundamentally mysterious substances like never before

There is no substance more plentiful and available to humanity than water. It is the easiest to study and yet remains one of the most mysterious. Understanding the interconnectivity between air and water represents an important piece of Nature's puzzle.

It is a known fact that the physical and chemical properties of water differ from all other chemical compounds in the Universe. Man-made materials may exclude water, but Nature does not; Nature centers on water and makes it applicable to all things, and in this chapter you will find understanding by answering these fundamental questions:

Why do no other compounds act like water.

Why do all organic creatures need water to live?

Why is water the most abundant substance in our bodies and in the Universe?

Even though every science textbook claims or infers that two hydrogen atoms (H_2) and a single oxygen atom (O) combine in Nature to create water, where do we actually see that happening? This fundamental question is rarely if ever asked, instead, we are just told that it happens:

Where can we observe Nature combining two hydrogen and one oxygen atoms to form water?

The answer is, nowhere that we know of. This should not come as a surprise. After all, scientists admit that water is still a great mystery. Each time modern science uses the terms mystery or anomaly, we understand that the current

ideas on the topic are probably incorrect.

A big red flag should have been raised when every chemistry book said H_2 + O equaled water, but no one had attempted to demonstrate how that happened naturally. Modern science learned long ago that every cell in every creature requires water to exist and to replicate. The most common chemical formula in the world is H_2O and the insight that water may not randomly combine in Nature was an essential ingredient for UM researchers that led to many other discoveries.

Many fundamental questions formed during the research period of air and water, such as:

Water is heavier than air, so why do clouds float?

This surprisingly simple question has no clear or correct answer in modern science. If you search the internet, you will find scientists trying to answer the question by relying on air updrafts. However, many cumulous clouds have flat bottoms, which clearly indicate that they are not being held up by wind drafts. Additionally, as any keenly observant person has witnessed; dust particles eventually *all* settle to the floor when air movement stops. Our atmosphere is no different, the wind is not always blowing, and if upward drafts were holding clouds in the sky, when the draft ends, the clouds should fall, but nobody has seen that happen.

Besides answering these fundamental questions and others, this chapter introduces many new natural laws and extraordinary claims regarding air and water. What happens in the atmosphere around us is concealed by chemistry and physics pseudotheories. Because you cannot see the air molecules, we have relied on modern science's theories, and they do not make sense. What does make sense is that all natural matter is organized, and that is apparent in all of the UM chapters.

The Chemical Model - UM Chapter 24

Chemistry studies the behavior, composition, and properties of matter. Similar to physics, chemistry is more abstract because chemists have traditionally struggled to observe and describe the chemical compound interaction on a small level. The abstract nature of chemistry is one reason many normal people think physics and chemistry do not really apply to them.

However, every time you wash your hair or eat something, it is important that the chemicals you put on or in your body are properly combined or they can harm you. The way elements and chemicals combine can mean life or death to most things living today.

Most historians believe salt helped the Egyptians gain power; the Roman Empire was built on bronze, the Spanish on gold, and the British on iron and coal. Just as these ancient compounds helped those who ruled the world in ancient days, we learn that 20th century uranium and plutonium brought WWII in the Pacific to an end. We can truly say that elements have power, and the elements run the energy producing plants that light our world at night.

The Big Bang and the pseudotheory that all elements came from hydrogen, forms a modern science creation myth. The theorists who made up these pseudotheories expect us to imagine or to accept the assumptions that all the important things happened 4.5 billion years ago.

You must remember as you read the Chemical Model that modern chemistry supposes that life randomly sprung from inanimate matter (nothing) long ago eventually evolving into you. Chemists wish to take credit for Life's beginning, blaming it on chemistry but the problem is, no chemist or any other researcher has ever observed life beginning from chemicals. We cover this in detail in this UM chapter.

Chemists mix several chemicals together and tell us what the results will be. They know because they have done it before. However, modern chemistry in its most basic form cannot do what it was designed to do, that is, describe and explain Nature's chemical processes.

For over 60 years we've taught the Element Pseudotheory in the classrooms and we learned that the elements in the periodic table are 'natural elements.' This came partly from the theorist's false claim that elements came from the Big Bang. Not only is the Big Bang unsubstantiated, the existence of pure elements in Nature has never been demonstrated.

The reason for this is that 'elements' are not natural substances and must be separated from natural molecules using technological processes to reach their pure state. In the Mass Weight Model chapter, we learn new evidence that explains why matter's weight changes; in the Chemical Model we learn why we must revisit the idea of elements' atomic weight. If all the atomic weight figures of the elements that make up the periodic table are inaccurate, imagine what changes might occur in chemistry.

Learning that all of the 'elements' on the Periodic Table are artificial and manmade, and realizing that they are not found in Nature in the form depicted on the table, we ask these fundamental chemical questions:

Where has science ever proven that Nature uses the elements in their pure form to create minerals?

How could Nature use pure elements if pure elements don't exist in Nature? What else does Nature use?

While humanity can use technology to break down natural chemical compounds such as salt into its artificial constituent elements Na and Cl, this technology does not prove that Nature does the reverse to form the chemical compounds found in Nature.

Salt is one of those topics that comes up repeatedly in the Universal Model because it has such universal importance, and because it is critical to all organisms. It affects all of our lives and more scientific fields of study than most people imagine!

Chemistry textbooks all show that Na + Cl = NaCl, or common table salt. Photos from school textbooks attempt to demonstrate how salt is made by showing sodium metal and chlorine gas combining in a fire-like reaction.

What is so amazing about this supposed claim is that chemistry students are told the formula is real and they are shown photographs and videos that presumably illustrate an experiment. However, as you will see, the photographs are literally a hoax and have been deceiving students for years.

Why is the experiment a hoax? Because the bright flash in the photo does not happen when Na (sodium) and Cl (chlorine) are combined at room temperature, no salt forms in this way. The magic happens only by dropping a small amount of water on the sodium, then BOOM! The Flash happens because sodium reacts violently with water, but that tidbit does not appear in the

Na + Cl $\neq$ NaCl (Salt)

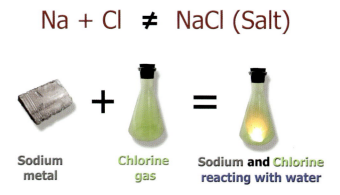

| Sodium metal | Chlorine gas | Sodium **and** Chlorine reacting with water |

textbook description. The chemistry textbook writers merely staged the photographs to make it look like a flash happened just as sodium and chlorine are combined.

In the Rock Cycle Pseudotheory, you will discover the origin of salt and salt deposits from around the world, which until the UM, have not been known. Geologists learn salt chemistry using textbooks that claim sodium and chloride ions combine on their way from the mountains to the sea and make salt, although no one has ever witnessed this process.

The experiments in this chapter demonstrate that the carbon dioxide in the air is *not* where most organisms, especially plants, derive their carbon. If plants actually make sugars by processing carbon dioxide and water, then a scientist should be able to reproduce this *simple* process, but why can't they? Because this is not the way sugar is made. Plants do not form their sugar by simply combining carbon dioxide, water, and light; just as water is not created by combining hydrogen and oxygen, or salt by combining sodium and chlorine. There are pieces of Nature's puzzle missing from the modern science understanding, evident because they cannot create these molecules with the constituent pieces they insist make them.

Extraordinary claims such as this require extraordinary evidences and the evidence for these claims, along with new natural laws fill the pages of this amazing Chemical Model chapter.

The Universe Model - UM Chapter 25

Before this chapter, we discussed micro-universe topics, of which Nature abounds. Our understanding of Nature is only as good as our understanding of these concepts. This chapter covers the larger, macro-universe topics that relate to our world from the upper atmosphere and out into the Universe itself.

For thousands of years, humans admired and studied the heavens using no other instrument than their eyes. It was not until Galileo famously turned his telescope toward the night sky that humankind began to see the heavens differently. Eventually a new science was born and cosmology with its new heavenly observations came along with a bag of new theories to speculate how and when the Universe appeared.

Cosmology includes the study of the Universe, and modern cosmologists like to think that they know both the origin and eventual fate of the Universe, however, as you will discover, their origin and fate of the Universe theories amount to little more than scientific imagination. The UM actually provides a correct description and explanation of the Universe and how it is currently organized.

This chapter also discusses astronomy, which is the branch of science concerned with celestial objects such as planets, moons, stars, and galaxies. 'Astro' from the Greek word 'star', and 'nomy' comes from the Greek word 'nomos' which means 'law.' Therefore, astro-onomy literally means, 'star laws' and as such, it should reflect the study of natural law in the heavens. Today, modern astronomy consists mainly of theories as this chapter demonstrates.

The size of the Universe is an important topic for astronomers as well as philosophers. Our relationship to the heavens and the organization of the Universe plays a critical role in our understanding about the relationship we share with Nature.

So how big is the Universe?

Although modern cosmology claims there is an end to the Universe, each time we build a bigger telescope we see galaxies farther away than before. How can we expect to answer how something formed if we don't even know what it is? Modern cosmology proves the fallacy of trying to learn without following the correct Learning Process. If you do not have the correct knowledge to start with—in this case, what the Universe is—then you will never comprehend the wisdom about how the Universe formed.

Modern physicists' holy grail is to develop a theory of everything, which means a tidy combination of the Relativity and Quantum Mechanics Pseudo-theories. While this has not worked out so well, it has not stopped well-known theoretical physicists from reusing the same old antidote of creating another theory, a tactic of modern science ever since Einstein. After several years, when the latest theory does not match reality, science either devises a new theory or implies that reality does not exist.

We know one of these absurd theories as the Big Bang Theory. Scientists claim they have settled this theory, at least in their mind, and they continue to teach it as such. Have scientists actually confirmed the essential parts of the theory? What range of experimentation and detailed physical observations did they use? How does one observe or verify that something happened 4.5 billion years in the past? Without observation, it is not science.

This chapter covers many new discoveries related to our universe, however UM researchers were also able to reveal new discoveries involving celestial bodies that you can see with your own eyes, such as other planets, moons, and our Sun. In this chapter, you will learn about a new law of planetary distance and other new incredible truths about our Sun.

All matter in the Universe would cease to exist as we know it if energy did not continue to flow through the matter. Like an electrical motor, remove electricity and it stops turning. Or the icemaker in a refrigerator; without a continual supply of water and electrical energy to the refrigerator, no ice would form. All atomic systems follow natural energy laws including our solar system and the surrounding Universe systems. But each one of these systems would fail if energy were not being fed continually into each system. This prompted the asking of a life altering fundamental question:

Where is all of this energy coming from?

This mind blowing fundamental answer talked about throughout the Universe System chapters completely changes how you will view our Universe.

Every time mankind has tried to limit Nature we have failed. Einstein said nothing could go faster than the speed of light. He was proved wrong by technology. Heisenberg said we could not know where and when electrons exist. He was proven wrong by technology. Today, cosmologists say that their telescopes are so powerful that they can see to the very edge of the Universe. Technology is proving them wrong again. Every time we look with a more powerful telescope, we have been able to observe far distant galaxies in a larger Universe.

Our technology has clearly advanced; however, our science has not. This is another clear example of how the Dark Age of Modern Science has blinded astronomers into believing their Big Bang theory; that the Universe has an end because of the beginning they imagine. Where is the physical evidence that the Universe has an end? The more powerful the telescope, the more endless the Universe seems.

The Meteor Model - UM Chapter 26

How many of us have had the opportunity to look up into the darkness of the night-time sky to see a shooting star enlighten our view of the Universe? What a life-changing, memorable experience this can be! When we see our first flash of light across the starry background, wonder fills our souls. Thanks to questioning everything with an open mind, everything in science, including meteors, is on the table for reevaluation. This brings us to this fundamental question:

What do astronomers really know about shooting stars?

My father tells a story about an experience he had one evening in the late 1990s. He was listening to a local radio station that had a guest astronomer on with the talk host. What caught his attention was a discussion about meteors. One caller had asked questions about these streaks in the sky; the astronomer answered him, explaining that the meteors we see on a typical night that produce the characteristic flashes of light hundreds of miles away are rocks the size of a grain of sand. The surprised caller said that he could not believe that such was true, but the astronomer assured him that it was, in fact, the size of a grain of sand. Both the host and caller expressed thoughts that it was just so hard to believe. As tension began to rise, the astronomer did his best to assure the caller that modern scientists do in fact know what meteors are.

It was at this point that my father could no longer resist and decided to call the station. A quick pick up by the call screener who queued up almost immediately. He began by mentioning he had some questions about meteors and meteorites and started right in with some fundamental UM questions addressed to the guest astronomer.

Most of talk show's science guests are prepared to handle textbook questions and answers, but you cannot find many UM questions in today's textbooks, so you can imagine the stumbling over the answers that occurred. After a couple of attempts, the astronomer finally responded with "Well you know, there are a lot of questions in astronomy we just don't know the answers to." And with that queue, my father said to the host, "I think that is the best answer we

have heard all night!"

Supposedly, when a meteorite passes through the Earth's atmosphere it forms a streak of light across the sky caused when the object burns up during reentry. Burning up due to atmospheric friction, modern science calls the light a meteor, or as it is known to the public, as a shooting star. We found no previous work that scientifically questions the prevailing meteor theory as presented by modern science.

In this chapter all the light streaks in the atmosphere are referred to as 'meteors', and many fundamental questions are asked challenging the validity of meteors coming from meteorites the size of a grain of sand.

Modern science teaches that small particles vaporize at 50-70 miles above the surface of the Earth because of air friction, leaving a flash of light. The origin of meteors as well as meteorites and craters are universally accepted across all fields of science today. The fundamental question that began the Meteor Model chapter is:

What if the origin and definition for meteors is incorrect?

Besides just causing streaks in the sky, meteors and their existence play a vital role in modern science's theories today. These theories are literally impacted by or based on rocks falling out of the sky.

Supposedly we get meteors (shooting stars) from meteorites, which are the rocks that seem to fall ever so often to the Earth. It is from these rock fragments that researchers often try to determine the Earth's age.

Meteorites come in many forms and supposed origins, the most common being tied age wise to the beginning of our Solar System, which, according to modern science, happened about 5 billion years ago. However, if the so-called meteorite that science 'dated' to 4.5 billion years old never actually came from space, the whole meteorite dating scheme falls apart. In addition, if the dating method used to date these meteorite rocks was never valid in the first place, meteorites lose their place in the geological dating plan. Both of these extraordinary claims find their evidence in the UM.

The principle of questioning everything with an open mind led to a very long list of fundamental questions about meteors. They bring a whole new understanding of meteors and how little modern science actually understands about them.

Because meteors are not being reported by radar, scientists say that they are only as big as a grain of sand.

When was the last time you saw anything the size of a grain of sand miles away be as bright as a meteor?

Radar and other detectors track over 10,000 man-made objects 10 cm and larger every day as they have for decades. It is possible to detect an object as small as 2 mm.

If meteorites fall to earth in the numbers estimated by astronomers, then why does radar not detect all of their paths?

We not only see shooting stars, we can hear them. Careful tuning of a radio picks up a blip coincident with the shooting star.

How does a rock falling to the ground from space generate a radio signal?

If meteorites and impact craters are actually connected in the high count modern science suggests:

Why has there never been one single confirmed, recorded instance in a scientific journal of a human being seeing, with their own eyes, a very, very hot meteorite fall out of the sky and form a crater?

Once you read this chapter and discover the true origin of meteors, there is only one answer to all the above questions and it is very simple. The primary reason for our misunderstanding about meteors is that modern science is unaware of their true origin. This final Universe System chapter will reconnect the light flashes we see in the sky to the Meteor Model found in the Hydroplanet Model, Chapter 7. The meteorites must have an origin and the place most come from is not far from home. The Meteor Model is also associated with the Weather Model and the combined information from these two chapters will certainly change the way you view the night sky, especially when you see a shooting star.

Reading the Universe System sheds new light on all matter. Simple answers to some of the Universe's biggest questions give new meaning and a new foundation affecting science fields associated with our Universe. Gaining a true understanding about essential measurements and by removing pseudo-theories, you can view the Universe in a whole new way. The new models in this chapter are just as revolutionary as when science learned the Earth was not flat, or when we learned the Earth revolved around the Sun. Built on observable evidence, not imaginary pseudotheories, these easy to grasp models make gaining true answers to the Universe's basic questions a joy.

THE MOST POWERFUL SCIENTIFIC
DISCOVERIES ARE THE SIMPLEST

Chapter 7
A New Millennial Science
Revolution - Reaction - Advancement - Education

*"The replacement of false theory with scientific truth
will be the foundation of Millennial Science,
a new science destined to last for a thousand years."*

Universal Model, Subchapter 4.4

The UM's effort to pioneer a new Millennial Science to carry humanity through the current millennium to heights never before imagined catalyzes the need for a scientific revolution focused on the learning and growth of true science.

Why must there be a scientific revolution?

In order to change the entire fabric of modern science by replacing its very foundations will require a scientific revolution, and the UM had to set that in motion. The building of such a revolution invokes many different reactions from both the public and scientists.

After two decades of conversations with those on both sides, it quickly became apparent who would be interested in the UM; not scientists generally, but the average individual who simply wants to know how Nature works and is willing to learn the new scientific truths found in the Universal Model and help in restoring true science to the world.

Before any revolution can occur there must first be a crisis. Over the last century, we have lived in a Dark Age of Science where false theories are being taught as fact all over the world. To facilitate this true science revolution, new discoveries must be assimilated into mainstream science with the previous false scientific views replaced with new scientific models. These new models must be able to account for and explain natural phenomena better than previous theories or models, which the UM claims to do.

What has been or will be the expected reaction to the UM?

Over the years, as various parts of the UM were being written, some individuals had been given the opportunity to pre-read some of its chapters. Their feedback has been important in helping develop various UM models and principles. Furthermore, we have had many opportunities to speak with groups of people in UM conferences and classes where we were able to share parts of the Universal Model with them. These people were very excited to learn more about a science that was refreshingly new. They expressed their feelings and thoughts in such a way that it quickly became apparent that mankind instinctively just wants to know the truth.

Along with public's reaction to the UM, there has been and will continue to be plenty of differing reactions from the scientists themselves. The basic characteristic of science should be the willingness to abandon a currently accepted theory when a new, better one is proposed.

For a scientist, it can be extremely difficult and even dangerous to venture outside the current scientific view. If one oversteps the bounds of mainstream scientific theory and practice, he or she does so at the peril of having a career derailed. Even if a few of one's peers can be convinced that a new idea or theory is correct, there is no assurance the daring scientist will not be looked down upon, ridiculed or barred from future opportunities.

The scientist who is looking for scientific truth will of course find the new truths in the UM exciting. Reactions from individual scientists have been and will continue to be varied. One of the wonderful opportunities this UM journey has brought to all involved is a deep respect for past scientists as well as today's scientists. Until one has tried to do what a scientist does, we cannot fully appreciate the hard work and dedication scientists commit to.

This magnitude of new scientific discovery has never been presented to the public before, therefore it is impossible for anyone to fully comprehend the magnitude of the reaction scientists, or people in general, will have towards the UM. From this we surmise that if, in general, the UM is what it claims to be—a new science for the next millennium—the reaction will be like nothing we could ever have imagined.

How will Millennial Science be advanced?

People in general today seem to find science either too boring or too confusing. Some seem to even fear it or think it's a made-up fairy tale. Who can blame them? How can the public discern between real science and science fiction when they are told by real scientists that they came from nothing and you have to just imagine most of their theories?

The Universal Model is a simple model; it is appealing, successful, and definitive. The individual models that make up the UM are also extremely cross compatible, simply building one after another. The experiments within the models are for the most part, simple to recreate and replicate the results because they are true.

Anyone can teach the simple principles outlined in the UM. After only a few months of reading the UM in their spare time, those who were not professional scientists, began teaching the Universal Model to others who had never before heard of this new science. For the first time in history, true science can be seen by the general public as simple, fun and exciting to learn. This is the key to Millennial Science advancement, for science to be admired and loved by almost everyone.

With Millennial Science, anyone can practice basic science; everyone can understand and enjoy it because it makes sense. Natural science is no longer restricted to those who do research on abstract concepts. With the internet anyone can contribute to science today. Wherever there is a desire to learn more about Nature, Millennial Science will be there; Nature's puzzle has no edges, which means the future discovery of new truths is endless if we follow Nature's true principles.

Now that we have the Universal Model, how can we use it and how do we get pseudotheories out of our schools?

Whether we are aware of it or not, we are all teachers. We all have family or friends who we influence and teach by our actions. Our teachings may be good or bad, but everyone influences the world in one way or at one time.

Modern education is not far behind modern science in its deficiency to teach and show future generations true science. However, this weight does not fall on the teachers themselves as they are only following the directives and curriculum put into place by those with agendas other than teaching truth. With that said we recognize that there are many excellent educators who do the best they can every day with what they are given.

Science leaders and educators have a scientific obligation to teach the rising generation scientific truth and the public has a right to require them to do so. Reading the UM provides the base from which to require that we teach natural laws instead of pseudotheory in schools. With the citizens providing the funds for most of the research, and by electing leaders who push for teaching true science in public schools, Millennial Science will *gradually* replace modern science.

What will Millennial Science teach? Scientific truth. However, because so much of modern science has left truth out of its purpose, current education needs to restore the idea. This leads us to answer the question as to how we will remove the pseudotheories. We must spark the flame that will create a revolution in education in order for a New Millennial Education to be established.

Millennial Education will follow the general public's awareness of the Dark Age of Science. We must teach the new Learning Process and the Universal Scientific Method (USM) outlined in the beginning UM chapters. As the experiments and observations in the UM are verified by the public and scientists, the importance of the Universal Scientific Method will become apparent. The USM was the means by which the new understandings of Nature came to be in the UM. Accordingly, why not teach students the USM so that new natural laws can continue to be discovered?

Each of the USM's six steps and their importance should be taught to every student so that they can personally experience scientific discovery. As more students envision natural science in an orderly way instead of everything coming from nothing, it is anticipated that the subject of science itself will become more popular and new true scientific advancements and natural laws will be able to come forth.

Chapter 8
UM Extraordinary Claims & Discoveries
Over 600 & Counting

"Personally, the author found that every discovery in the UM was the most important as it was discovered."

Universal Model, Subchapter 27.4

When an extraordinary claim or discovery is made in just one field, it often affects the whole of science. The Universal Model makes hundreds of extraordinary claims over many differing fields of science. UM Extraordinaries are scientific claims, discoveries and evidences that are so important they will replace the current science taught in textbooks.

They must also be scientifically verified, for extraordinary claims require extraordinary evidence. This is a common statement in science and one used frequently in the UM. The UM is unique because of the number of extraordinary scientific claims and evidences it contains.

As you read the following partial list of UM extraordinary claims and discoveries, remember that each is backed by extraordinary evidence discovered in Nature and described in the hundreds of pages in the Universal Model. Never before has there been such a compilation of new discoveries in science. Some may think there are no additional significant discoveries left to make; this is far from true. Some of the greatest discoveries are yet to be made, once a true foundation can be established.

20 UM Extraordinary Claims & Discoveries:

1. The entire world has been in a Scientific Dark Age for more than a century during which not even one significant new natural law has been discovered.

2. The idea that our Earth is not a magmaplanet, and that the actual existence of magma is a pseudotheory.

3. The modern-science Rock Cycle is a pseudotheory.

4. Many mysteries exist in geology today because the true origin of rocks is not understood.

5. Major catastrophic events occurred in the past which affected and shaped the earth in ways never observed by modern science.

6. More scientific evidence exists about the Universal Flood than any other global geologic event in Earth's history.

7. The Hydroplanet Model explains the origin of the volume of water necessary for the Universal Flood.

8. Modern Science does not understand the true origin of weather, including the true origin of lightning, auroras and the Earth's energy field.

9. The modern science dating system is fatally flawed and its faulty dates must be replaced with verifiable dates before we can understand when natural events actually occurred.

10. Because the Earth's age has been based on the dating of allegedly magmatic (igneous) rocks, the Earth's age estimates are incorrect.

11. Active, in-process fossils do not exist anywhere in Nature.

12. The fossil layers that supposedly succeed one another in the same evolutionary order do not exist anywhere in Nature.

13. The Evolution Pseudotheory is a belief, not science, which denies natural laws and mingles truth with error.

14. Without measurement, there can be no science and some of the most fundamental measurements in modern science are incorrect.

15. Modern science does not have the correct definitions for mass, weight, length, time, energy or matter.

16. The Theory of Relativity is a pseudotheory.

17. The true nature of the atom and its structure is delineated.

18. The origin of the dual nature of light is explained for the first time.

19. A new Revolutionary Universe Model is shown to replace the Big Bang Pseudotheory which has contributed to the current Scientific Dark Age.

20. The true origin of most meteors, meteorites, and craters are set forth and demonstrated to be different than currently taught in modern science.

As we end this Summary of the Universal Model, it is my hope that your introduction to the UM has been a journey worth pursuing. The new discoveries found within the UM, which are true, will have a lasting impact on everyone in society whether they choose to learn them now or later. No one can escape the progress of society as these truths become integrated with new technologies that continue to help us understand and be able to explain the world around us in ways never before imagined.

Discover the Universal Model at:

UniversalModel.com

One Can

One closed mind can lead to a century of darkness.
One pseudotheory can destroy the very foundation of science.
One deceitful researcher can lead an entire country into error.
One crater can misrepresent the true age of the Earth.
One measurement of error can greatly modify the Earth's mass.
One concealed ray of truth can send all of cosmology into darkness.
One piece of discarded evidence can impede growth and answers.
One idolized false scientist can mislead an entire generation.
One agenda can negatively impact the discovery of truth.
One incorrect piece of information can alter how you view yourself.

One open mind can revolutionize scientific history.
One fundamental question can lead to new natural laws.
One flash of true inspiration can provide a lifetime of new discovery.
One insightful gaze into the past can reveal a glance of the future.
One newly discovered particle can transform physics and chemistry.
One drop of water can alter the definition of gravity.
One Universal Flood can correct all the Earth's geology.
One Clovis point can reveal the origin of man.
One correct answer can lead to an entirely new science.
One can restore true science with a Universal Model.

INDEX

Image Credits

Leonid Meteor Image by Navicor,
Creative Commons, 3.0
Most images were taken or created
by Dean. W. Sessions
Other images by Shutterstock